Holographic Scanning

Holographic Scanning

LEO BEISER

Leo Beiser Inc.
Flushing, New York

WILEY

John Wiley & Sons

New York / Chichester / Brisbane / Toronto / Singapore

Library of Congress Cataloging in Publication Data:

Beiser, Leo.
    Holographic scanning.

    (Wiley series in pure and applied optics)
    1. Scanning systems. 2. Holography
I. Title. II. Series.

TK7882.S3B45 1988   621.36   87-28001
ISBN 0-471-80612-9

Printed in the United States of America

10 9 8 7 6 5 4 3 2 1

To Edith,
Ronni and Steve

My Illumination

Preface

Why holographic scanning? Which holographic scanning? It's hard to say. But in this book, I have tried. Holographic scanning derives from two parent disciplines: holography and optical scanning. The combination was inevitable and the combinations are innumerable. I approached holographic scanning on behalf of the researcher and practitioner seeking information as to which form should be selected for further development for a particular task.

"For a particular task" is a key concept. The reader will soon conclude that there is no panacea. In relieving some major limitations manifest in conventional optical scanning, holographic scanning may introduce some of its own. A clear perception of the classes of holographic scanning and their common characteristics is invaluable to guide optimization. I believe the reader will appreciate this.

The reader is probably one who is studying or already working in the expanding field of information transfer—communicating with images and data. Optical scanning transforms images to data, stores or retrieves data, or transforms data back to images. What a powerful discipline! I wrote this book to provide clear options and to inspire new directions for exercising that discipline.

I organized the book to transport the reader through the basics and into the real world of operating systems with minimum diversion. Relevant material available elsewhere is extensively referenced. Time-consuming analyses or historical background are reserved for appendixes. Thus the technology is addressed expediently and grouped, categorized, and assessed in light of this prior orientation. I strove to consolidate this broadly distributed field, hoping that the discipline will aid the reader—the researcher and worker in whose hands rests the progress of holographic scanning.

Progress is a continuum. Although Appendix 1 offers a historical review, I would like to take a moment now to acknowledge some of the pivotal contributors to this technology. Dennis Gabor not only invented holography, but—as is generally not known—also researched holographic scanning. Ivan Cindrich published the first and one of the most dramatic statements on holographic scanning. McMahon at Sperry and Locke in Canada, and my colleagues at CBS Laboratories, McMann, Ih, Darcy, Kleinschmitt, and Dalisa, made significant contributions. With the inevitable thrust into business graphics, the legacy of Bob Pole of IBM often reappears. Charles Kramer while at Xerox viewed holographic scanning through new angle—as did Antipin and Kiselev of the USSR. The book identifies so many more. Gerbig not only innovated, but encouraged the writing of this book. Brygdahl, Lee, Case, Sincerbox, Dickson, Funato, Ikeda, Ishii, and Ono researched wide ranges of technology. All of the above, and more, are acknowledged for their valued contributions to holographic scanning. They inspired this writing.

Leo Beiser

Flushing, New York
January 1988

Contents

Holographic Scanning

1

Introduction

1.1. INFLUENCES AND DIRECTIONS

Holographic scanning is a dual factor in the phenomenon called the information explosion; it is both a *product* and a *contributor*. This new technology, derived from modern optics, renders images and data more quickly and more accurately. With the proper motivation and control as it matures, the field of holographic scanning can become one of increased operational utility.

The aim of this volume is to illuminate the path of holographic scanning through the very discipline it advances—information. In striving to consolidate and organize the pivotal information—from historical perspective to design optimization—this volume seeks to serve the researcher and worker in the field both to clarify the basics and to enhance the procedure that can enable holographic scanning to expand its contribution to information handling.

What is *holographic scanning*? Briefly, it is a method of controlling the direction of an optical beam by articulating a hologram in the path of the beam. In Chapter 2 we devote more attention to the definition of holographic scanning and how it is distinguished from other optical scanning techniques. At this introductory level, let it be clear that we are addressing an elegant form of optical scanning. What makes it elegant is not only its relative novelty but also its apparent simplicity, which can divert attention from the discipline necessary for its control. In composing this book, we strove to provide the reader with a cross section of the spectrum of options available and the controls needed to heighten utility for a particular task. This is because no single holographic scanning technique can serve all purposes equally well. That this is also true for the general field of optical scanning is, in fact, the reason for the existence of holographic scanning. A compilation of optical scanning literature (Bei 9)* indicates the diversity of this discipline.

*The reference list is located at the end of the book.

1.2. OPTICAL SCANNING

Optical scanning may be defined as the controlled sampling of spatial data. A beam of light is focused to a fine point and caused to move across a field of data in a predetermined pattern. For the purpose of information handling, two complementary scanning processes are available.

In one, the scanning beam illuminates an object in a prearranged series of positions. At each point along its path, the light beam is modified by the object, forming changes in its intensity and/or direction. (Other modifications, such as polarization changes, can also be utilized.) A photosensor is arranged to view the object so that the changes are detected as a corresponding sequence of electrical current changes. This forms the signal, which represents a signature of the original object. This process is called, quite descriptively, "flying-spot scanning," stemming from the earliest work in television sensing and represented in this context in the historical review in Appendix 1. A contemporary flying-spot scanner is the cathode-ray tube (CRT), whose deflected electron beam is incident on a phosphor that generates a scanned point of light imaged by a lens on the object. Higher speed and higher resolution are available with a scanned laser beam. One of the methods of accomplishing controlled positioning of a laser beam—holographic scanning—is the subject of this book. Other descriptions for flying-spot scanning (or simply, "scanning") are "image capture" and "digitizing" (if actually digitized in position and intensity). Seldom is the light beam caused to increment over a set of discrete positions along the scanned path. The scanning process is usually continuous, convolving a point-spread function across an object space to yield an analog signal that may be modified subsequently by analog-to-digital conversion.

In the second process, the field of data exists in electrical form or in computer memory as a patterned signal, derived, for example, from the process described above. It is now required to construct the corresponding spatial pattern (image). Again, a beam of light is scanned. But this time, it is projected on a photorecording medium such that it exposes the medium over a sequence of positions. This sequence (scan pattern) must correspond to a prearranged plan (be synchronized) and the light beam must now be modulated (usually in intensity) to correspond to the changing current which represents the desired image. This process is usually called "recording." In the field of business graphics and graphic arts, it is sometimes called "printing." But it does not become a "print" until something more is done to make the latent image visible, such as toning or inking.

Again, a popular contemporary source of scanned light for recording is the CRT, benefiting not only from its very responsive beam deflection capability, but also from its equally effective facility for beam intensity modulation. And again, more recently, recording has been conducted with the scanned (and modulated) laser beam.

The processes of scanning and recording are combined in some systems, most notably, in facsimile transmission. A document is placed in the scanner portion of the instrument, which translates the original image into a corresponding electrical signal. This signal undergoes a series of operations called "image processing," which can include digitizing, bandwidth reduction, and image enhancement. The resulting signal is transmitted via one or more communication channels and finally, presented to a companion facsimile machine in another location. There the signal is reconstituted and presented to a recorder portion, which reconstructs an image that is to duplicate, as closely as practicable, the original document.

Facsimile is a generic system that may be recognized in a variety of forms, including electronic mail, intelligent printers and copiers, electronic publishing, and photoreconnaissance. Some of the most advanced of these systems, which scan and record the image using laser beams, excel in the combination of high resolution and high speed. The spectral purity and low noise of the laser provide for focusing to a fine point with controlled intensity and sufficient power to override systematic noise. Since its focused spot size is ideally not a function of its intensity, resolution (which depends on small spot size) and speed (which depends on energy per exposure time interval) are, in the first order, independent variables. Consequently, the field of laser scanners and recorders has had a dramatic growth, following almost immediately the availability of the practical laser. Further historical perspective is provided in Appendix 1.

The field of laser scanners has undergone its own period of maturation. Of the approximately 40 techniques researched and developed (Bei 6), four principal types have survived (Bei 9): the rotating polygon, the acoustooptic deflector, the galvanometer (broadband and resonant types), and the holographic deflector. The literature is well endowed with descriptions of the first three. While several fine reviews of the holographic scanner have appeared (Dic 3, Ger 4, Ger 5, Sin), none has attempted a comprehensive exposition of the field. The purpose of this book is to help fill that void.

1.3. COMBINED DISCIPLINES

Holographic scanning is a combination of holography and optical scanning, two independently complex disciplines. The combination of the two provides such unique and diverse characteristics that this book is dedicated to its organization and clarification. Little space can be allocated, however, to the fundamental underlying subjects of holography and optical scanning. Thus the reader need be well oriented in both prerequisites. In the recurring examples of this joint utility, significant referencing is provided throughout for acquisition of this background information.

Holographic scanning results from the articulation of holograms within an illuminating field: some by rotation alone, some by translation alone, but usually by combinations of the two. Thus the latter, which appears as rotation of a member about an axis, occupies the main body of this work.

1.4. BENEFITS AND LIMITATIONS

Why scan with holograms? There are several reasons. Some applications have materialized to provide a firm cause-and-effect relationship between the unique properties of a selected holographic scanner and its derived performance. Complementing the potential benefits, there are sobering influences which relate primarily to the burdens of creating the devices cost-effectively with sufficient accuracy and repeatability. The following comments offer an appraisal of these factors.

1.4.1. Virtue of the Radially Symmetric Substrate

Some benefits derive from deposition of the hologram on the smooth surface of a substrate exhibiting radial symmetry; (i.e., disk, cylinder, sphere, cone, etc.). When such a substrate is rotated about its axis of symmetry, several properties ensue:

1. Aerodynamic loading and windage (Mar, Len) is reduced significantly during high-speed rotation (with the elimination of polygon-like facets which act as paddle wheels). This reduces correspondingly the rotary drive power requirements, the environmental aerosol turbulence, and the consequent acoustic noise. It is well appreciated that the typical polygon measuring, say, 100 mm across its flats may not be driven at speeds exceeding approximately 25,000 rpm without requiring excessive drive power and operation in a reduced air (or inert gas) enclosure. In contrast, the Holofacet scanner (Section 4.2.1.1), which has a 100-mm-diameter spherical rotor, has operated at 52,000 rpm *in air at normal pressure* to attain thus-far unmatched performance.

2. Differential inertial deformation (Bei 6) is reduced, again due to rotation of a homogeneous substrate material having no radial differential expansion. The prior example of very high speed operation formed a 5-μm scanned focal point from an $f/6$ optical cone—a consequence of this added surface stability, which becomes even more important operating in reflection (per Section 5.3.2).

1.4.2. Reduced Wobble Near Bragg Transmission

In addition to the added dynamic stabilities noted above, there is, under certain conditions, increased tolerance to positional and surface nonuniformities. These are accrued when operating in transmission at or near the Bragg diffraction regime. Operating at the Bragg condition (see Section 4.2.2.4) reduces significantly a sensitivity to substrate wobble and operating in transmission (Section 5.3.2) also reduces the effect of surface irregularity.

1.4.3. Fabrication Accuracy Advantages

Further potential for increased accuracy is manifest in the manner of formation of the holograms, as compared to the generation of facets on a polygon. The accuracy addressed here is analogous to the facet-to-facet angular integrity which tends toward equal intervals between successive scans; that is, periodicity in the *along-scan* direction. Also considered is the instability due to wobble noted above, the equivalent of facet-to-axis and bearing runout nonuniformities in the quadrature *cross-scan* direction. The procedure selected during holographic exposure can, under stringent control, benefit both forms of equivalent facet nonuniformities. This derives from the in situ and noncontacting process which may be implemented in formation of the ''master'' holographic scanner. The substrate is mounted on a spindle connected to an angular indexing device which provides angular accuracies that exceed the operating accuracies required. In addition, the optical components must be assembled with commensurate stabilities for the reference and object beams. Then, by successive exposure of the sensitized substrate between precise increments of angular indexing, latent holograms are formed that represent idealized angular orientations—no mechanical contact on the substrate except through its indexed shaft. Clearly, the processing, remounting, and replication procedures (if any) need to be planned and executed with commensurate integrity. It is further understood that the advantages of in situ and noncontacting exposure are effectively voided by piecewise assembly of a holographic scanner—either from interferometric or computer-generated masters—in which case, conventional procedures must be instituted to ensure angular and positional integrity. These brief considerations not only exemplify one of the potential advantages of holographic scanning but also heighten the need for discipline and instrument control to benefit from this technology.

1.4.4. Partial Benefits

The sometimes-invoked advantages of the unique options of the holograpic scanner—in angular magnification, multidimensional scan, or retrocollection—

must be viewed with care. While a range of angular magnification is realizable ($m = \Theta/\Phi =$ optical/mechanical angle ratio, interpreted in Section 2.8.4), it is also true that the pyramidal polygon is typically operated at $m = 1$, and a prismatic one at $m = 2$. Also, under the condition of focused illumination, a polygon can provide a range of m from 1 to 2. This relates to the parameters of the scanned limaçon (Bei 6). Figure 2.2a illustrates a prismatic polygon operating in the nonconventional mode of $m = 1$. In Section 4.2.1.3.2 the equivalency of a particular holographic scanner and some polygons is discussed. Other comparisons appear in Sections 2.8.4 (Bie 8) and 4.5.3.

With regard to multiplexing and multidimensional scan (covered in Section 4.4), there are very few conditions that benefit uniquely from the scanning of holographic elements as compared to scanning of conventional optical elements. The holographic elements may, under controlled conditions, be disposed more accurately, as discussed earlier. Some special computer and interferometrically generated scanners do provide beam movement in directions other than hologram movement, described in Sections 4.4.3.1 and 4.5. Also, some more conventional holographic scanners can benefit from the ease of forming unequal facet lengths having unequal focal lengths. These provide, for example, a pseudorandom field of scans for bar-code reading (Sin), discussed in Section 4.2.2.6.1.1.

Sometimes expressed as advantageous is the process of retrocollection, in which a portion of the flux scattered from the scanned object is collected by the holographic aperture to be descanned and returned along a stationary path for signal detection. This reciprocal process, discussed in Section 4.2.1.3.2 and A1.4, may be applied to *any* optical scanner having a large collection aperture. The principal advantage of holographic retrocollection is its automatic spectral filtering, which could otherwise require auxiliary filtering components to suppress unwanted background illumination.

1.4.5. Limitations

The need for caution and for diligent control of device fabrication and assembly merits reemphasis. The formation of holographic elements exhibiting near-diffraction limited properties over large apertures at high efficiency is a demanding task. Yet this is but one prerequisite for successful operation of a holographic scanner, which must convey the outputs from a succession of such elements with high uniformity and integrity. An index of the importance of these factors is manifest by their discussion throughout this book, notably in Chapter 4 and by the dedication of an entire chapter (Chapter 5) to scan integrity.

An intrinsic limitation of the diffractive element is its spectral selection sensitivity. We take advantage of this property far less frequently (Section 1.4.4) than we encounter it as an impediment. Wavelength shift between exposure and reconstruction is a problem for holograms exhibiting optical power or formed on nonplanar surfaces. Even when formed as plane linear gratings, there are disadvantages to use the same grating for anything but a single wavelength in reconstruction. This constrains multicolor scanning to diffraction from individual gratings, one for each color. Even then, unless the system is radially symmetric, it suffers from differential nonstraightness, nonlinearity, and inequality of scan angle, endangering color registration.

Radial symmetry can provide extremely wide-angle uniformity and integrity. However, some systems depend on departure from radial symmetry for the advantage of, for example, wobble reduction, in which case they are exposed to complex interaction for the balancing of scan bow, linearity, and diffraction efficiency. Because these factors are generally compensated by superposition of complementary functions over a useful operating range, they are limited in scan angle. Fortunately, some systems can operate satisfactorily within this limited range.

Most systems require more rigorous compliance to scan linearity. When departing from radial symmetry, this imposes extra demands on the "flat-field lens" which transforms the angular scan of a collimated beam to a near-linear spot displacement in image space. The functional relationship between the spot position and deflection angle is therefore a variation on the conventional "f-θ" lens, requiring the special design and fabrication of lenses to match holographic scanners operating in this mode.

Finally, the classic quest for high diffraction efficiency is a constant parameter of concern in holographic scanning. This is manifested not only in absolute efficiency but also in uniformity from facet to facet and in its variation over each scan interval. Although notable progress in these parameters is apparent, the most effective diffractive designs can fall short of efficiencies and uniformities available from conventional alternatives, and sometimes require adaptation of closed-loop correction techniques (Cat).

1.5. ORGANIZATION OF THIS VOLUME

This book has four additional chapters and closes with three appendixes; so organized to retain continuity and to provide for independent reference.

1.5.1. Chapters

In Chapter 2 we express the characteristics of holographic scanning, placing them in perspective within the general field of optical scanning. A recurrent theme is the basic similarity of holographic scanning to classic optomechanical methods. Special attention is devoted to the properties of scanned resolution.

The parametric relationships are developed progressively, yielding general equations which include such factors as scan magnification and duty cycle, as provided by holographic elements which can be displaced from the rotating axis and launch collimated or focusing output beams.

Chapter 3 covers some of the fundamentals of holography. The zone lens appears not only as a classic prototype to holographic lenses, but as an often-invoked analog to a range of types and properties of scanned holographic optical elements. The interferometrically generated hologram is reviewed at the close of this chapter—one that elicits repeated reference throughout the book for both interferometric and computer-generated holographic scanners.

The central mass of this work is represented in Chapter 4, covering holographic scanning techniques. The chapter is divided into five sections, as follows. Section 4.1, in which we discuss illumination-boundary constraints, introduces a new diagnostic tool called a conformal ray surface (CRS). This offers heuristic appreciation for the characteristics of radial symmetry and the consequences of departure from such symmetry. A variety of basic forms of rotating bodies are discussed in light of these consequences. Each represents the basis of subsequent discussion as holographic scanners.

Section 4.2, a description of rotational scanner development, is divided into two parts, depending on whether the substrate provides a surface component which is normal to its rotating axis. This determines whether the output beam may be launched normal to the axis, a criterion for the generation of straight-line scans. The nonnormal type is best represented by disklike scanners, which depend on various scan-straightening techniques to benefit from substrate simplicity.

In Section 4.3 we discuss translational scan, which provides useful and transferrable properties to rotational systems. Special narrow-aperture translational systems are described to provide (preprogrammed) arbitrary two-dimensional scan with low aberration.

Section 4.4 provides views of multiplexing and multidimensional scan. It highlights the difficulties of utilizing the same holographic facet for multiple purposes (introduced in Section 1.4.5) and offers a variety of techniques for combining functions in one, two, or three dimensions.

Section 4.5 discusses computer-generated hologram scanning. Following an introduction to the technology, it reviews the important process of conformal transformation to transfer a computer-generated hologram ''strip'' onto a rotational substrate, and closes with considerations related to system utility.

Chapter 5 covers the closing major subject of this work, mechanical-optical integrity. As reflected in the earlier remarks, it demonstrates, in part, the dedication and balance needed to construct a useful holographic scanner system. The chapter is divided into four sections, on the subjects of anamorphic correction, scan linearity, substrate errors, and wavelength-shift errors.

Although anamorphic error correction was originally developed for conventional scan techniques, it is equally useful for holographic scanning. Several novel analytic and application approaches are presented in Section 5.1.

The subject of linearity is accorded special attention in light of the operation of many deflectors in the nonradially symmetric mode. Some general equations are developed in Section 5.2 which describe this form of nonlinearity.

Substrate quality and stability factors are discussed and evaluated in Section 5.3, addressing both gross orientation errors and surface deformation consequences in reflection and transmission.

In Section 5.4 we provide an analysis and review of wavelength-shift errors and their possible correction by balancing techniques.

1.5.2. Appendixes

Appendix 1 is a historical review. It provides a perspective on the development of holographic scanning which reveals significant independence of innovation. Apparently, holographic scanning appeared as a natural outgrowth to workers who were operating in both holography and scanning in the mid-1960s. A surprising forerunner to the modern holographic scanner is revealed which appears to have frustrated some patent coverage.

Appendixes 2 and 3 exemplify this apparent independence of innovation. Appendix 2 is an annotated translation of work conducted in the USSR by M. V. Antipin and N. G. Kiselev. Appendix 3 is an annotated abstract from a U.S. patent by C. J. Kramer. Major attention was devoted to both to fill voids and to document them with rigor and completeness. While some of the conclusions expressed in each are similar, their approaches differ significantly. After a review, Appendix 1 offers the conclusion that these works were, indeed, independent.

1.6. SUMMARY

The principal thrust of this work is to heighten awareness of the options available and the discipline necessary for researching, improving, and practicing the art of holographic scanning. The motivation for its advancement is the untapped reward of this maturing technology, applied to the expanding field of information handling.

We concentrate on those options and disciplines that are unique to this technology. Others may be gleaned, when needed, from the extensive associated literature. An example is the important component of holographic recording media, which has benefited from long-exercised attention to the advancement of holographic optical elements (HOEs) (Cas 1, Cha 3, Cha 5, Che 2, Clo 2,

Eng, Mat 1, Soa). Noteworthy literature includes data on photoresist (Agm, Bar 2, Bar 3, Bee, Iwa 1, Joh 2, Nor), dichromates (Cha 4, Dic 3, Gra, Oli 1, Oli 2, Pen, Ral 1, Ral 2, Sol 2), silver halide (Coo, Fim), and general studies (Har, Joh 1, Kod, Kur, Loe, Moh, Smi, Wre, Yok, Zec).

We seek to construct timeless foundations for this technology, organize its disciplines, and correlate its evolution. But it continues to evolve as we write. With this writing, then, we offer a documentary of evolution, to serve as a base for progress in holographic scanning.

2

Characteristics of Holographic Scanning

2.1. A COMBINATION OF HOLOGRAPHY AND SCAN TECHNOLOGIES

The two technical prerequisites to holographic scanning are holography and scanning. While both have been documented as individual disciplines, we give them joint attention, to provide a secure understanding of their interrelationships and developmental routes and to offer a background for the exercise of creative skill by serious workers in the field. A comprehensive knowledge of holography and scanning is essential for advancement beyond the present state of the art. Superficial exposure to either can render one unjustifiably shielded from subtle complications that can inhibit the advancement and effective exploitation of this technology.

As discussed further in Sections 3.4 and 3.5, a hologram may be represented as an optical component exhibiting diffraction characteristics such that it transfers incident illumination from one direction into other controlled directions. If it did not invoke diffraction, this goal could be met completely by mirrors (plane or curved), prisms, and/or lenses. Indeed, analogies of a hologram for this application are those of a mirror (reflective) or a prism (transmissive), with or without optical power (lenticular), to converge or diverge the incident illumination upon a new location. This does not limit the interpretation of the "thin" or "thick" hologram; the diffractor in both cases is macroscopically thin. Holograms distributed within a substantial thickness (>100 wavelengths), used primarily for image multiplexing, are seldom applied to scanning.

An optical scanner may be represented as a device that *dynamically* alters the direction of nominally fixed incident illumination. This renders a quasi-continuous change in position of the output illumination as a function of time. It is the dynamic process that differentiates scanning from the static alterations of direction by passive components, including the hologram. Thus we combine the passive hologram (as the other fixed optical components have been) with a spatially active member to form a scanned output function of space and time.

A holographic scanner is, therefore, an assembly of one or more scanned holograms. Whereas the acoustooptic deflector (Bed) may be viewed as a diffraction grating having dynamically varying pitch which exhibits the characteristics of a variable thick (Bragg regime) hologram (Leg), holographic scanning typically entails mechanical orientation of one or more rigid holograms. Orientation is usually by rotation about an axis: hence the rotating polygon prototype. Although orientation by translation alone is feasible and will be discussed (in Section 4.3), unless otherwise indicated, subsequent discussion relates to the most prevalent scan utilization—rotation about an axis. As developed in Section 2.8, this rotation often includes an equivalent component of translation.

Although the principal task of the holographic scanner is to generate controlled output beam movement, it must usually sustain the following additional operational characteristics:

Provide an output function exhibiting

1. A high-quality wavefront—approaching stigmatic; free of aberration
2. Uniform output quality—independent of position ⎞ over many positions;
3. High optical efficiency—independent of position ⎠ high resolution
4. A plane scan function—nominally in the plane normal to the image surface
5. Options for field flattening and use of anamorphic optics (Section 5.1)
6. An allowance for wavelength shift—for reconstruction at a wavelength different from that used in construction of the holograms
7. Options for multiplexing and multidimensonal scan.

Be formed on a supporting substrate having

1. Simple surfaces—for fabrication ease
2. Replicable holograms—for production economy
3. Low wobble and centration perturbations—for synchronous and congruent scans
4. Low inertia and windage—for low drive power, noise, and mechanical perturbation
5. Options for reflective or transmissive operation.

No single technique will accommodate all of these objectives simultaneously. The task is to select and enhance the approach that maximizes the desired characteristics.

2.2. HOLOGRAPHIC SCANNER CHARACTERISTICS

A useful general classification scheme is by major contrasting characteristics (Bei 6), whereby laser scan technology is first divided into high-inertia and low-inertia techniques. High-inertia scanning is exemplified by the rotating components and the general rotational or translational orientation of relatively massive mirror, prism, or lens assemblies. *Low-inertia scanning* is exemplified by the nonmechanical techniques such as acoustooptic or electrooptic devices and includes the general class of specially designed low-mass translational (e.g., focus correcting) and rotational (e.g., galvanometer) components.

Since the holographic portions of scanners are disposed on a substrate that bears substantially all the inertial mass (as does the substrate for a mirrored surface, and even proportionately more than the substrate of a Fresnel lens bears with regard to its lenticular ridges), the dominant class of holographic scanner is the high-inertia device. This is particularly apparent with most holographic scanners having an array of holograms disposed on the near-periphery of a rotating substrate. As is well appreciated, the prime movers (e.g., motors) for the holographic scanners can contribute significant system inertia. Thus, as long as the scanner must be actuated (usually, rotated) the composite effect of the holographic scanner is that of a high-inertia (rotational-mechanical) system.

With the vibrational-mechanical scanner (e.g., galvanometer) assigned to the low-inertia category, there is no fundamental reason not to orient a hologram suitably on a low-inertia oscillating shaft—except that a useful purpose for doing so must be established. Value is to be gained by replacing a galvanometer mirror with a hologram. It will later become apparent (Section 4.2.2.4) that direct replacement with the shaft axis in the plane of the hologram yields no useful benefit. One may, however, take advantage of unique spectral or lenticular characteristics not normally available from standard components (Tai). In axial translation, one may take advantage of its lower mass for focus correction. Such holographic components fall into the class of holographic optical elements (HOEs). We need to confine our appraisal, however, to those holographic elements that are so configured and articulated as to provide spatial scanning.

These considerations do not preclude the possibility of the holographic scanner intruding into the application of the low-inertia devices, particularly those of the acoustooptic and galvanometer scanner (Bei 6), providing relatively uniform recurrent scans. While the strength of the acoustooptic scanner is based uniquely on its high-speed nonmechanical operation, the potential cost-effectiveness of the holographic scanner operating in this moderate-resolution domain (a maximum of approximately 2000 elements per scan in the conventional Bragg regime) and its capacity for providing multidimensional scan could motivate a shift in appraisal. Similarly, while the strength of the galvanometer is in its relatively low cost and slow-speed flexible scan, the potential cost and mass reductions of the holographic scanner could intrude on the galvanometer as well–again, where operation is in the relatively uniform recurrent scan mode. In the random-access or flexible scan modes, acoustooptic and galvanometer devices retain a clear advantage.

Deferring detailed discussion of resolution for subsequent analysis in Section 2.8, since a numeric value was expressed above, it is useful to clarify the term. As sustained throughout this book, resolution is taken as the total number N of diffraction-limited elemental spots of width δ (overlapping at their 50% intensity points) during the active scan interval (excluding blanking or retrace). The above-stated 2000 elements per scan therefore represents a continuous series of $N = 2000$ δ-sized elements having center-to-center spacings also equal to δ. Thus, if $\delta = 1$ μm, the total format width is $W = N\delta = 2$ mm; or if $\delta = 1$ mm, the total format width is 2 m. The key factor describing holographic scan resolution, as for all optical deflectors, is the value of N: the number of elements that occupy an *active* scan period (i.e., omitting the blanking or retrace interval). This value is invariant with subsequent passive optical magnification or demagnification which is stigmatic (aberration-free) and free of vigneting.

2.3. CLASSIFICATION OF HOLOGRAPHIC SCANNING SYSTEMS

Optical scanning systems have been classified (Bei 6) into three regions: preobjective, objective, and postobjective, ascribing distinctive characteristics to each. Although holographic scanning does not modify these classifications, some special interpretation is required. Figure 2.1 shows the designation of the scan regions as part of a general conjugate image transfer of a *fixed reference (object) point P_o to a moving focal (image) point P_i*. The reference point is affixed to

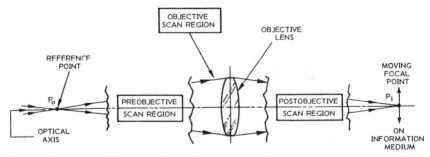

Figure 2.1. Conjugate imaging system showing scan regions that provide a moving focal point. Objective scan: lens translates with respect to information medium. Preobjective and postobjective scan: predominantly angular change. Angular and translational scan can occur simultaneously.

machine coordinates, as is, for example, the pinhole spatial filter in a prior beam expander; or the reference point may be effectively at minus infinity, serving as the origin of collimated light.

While other optical elements may be interposed, the lenticular element that provides final convergence of the wavefront to the moving image point is here defined as the objective lens. It may be a holographic optical element. Basic scan mechanisms are limited to two processes:

1. Translation with respect to the information medium. This includes translation of the lenticular element or translation of the medium, or both—and is identified as *objective scan.*
2. Angular change of the optical beam with respect to the information medium. This includes accessing the beam before or after the objective lens—identified as *preobjective scan* and *postobjective scan*, respectively.

Conventional angular scanning elements are, themselves, usually free of optical power, exemplified by the use of plane mirrors on rotating polygons and galvanometers, and (almost) linear gratings in acoustooptic deflectors. The compound acoustooptical scanning technique known as "chirp" or "travelling lens" (Bed) falls into the category of translational objective scan. Although it is possible to add lenticular power to a mirrored angular scanner, it is seldom so configured. The scanner is either before (preobjective) or after (postobjective) a fixed objective lens group. In holographic scanning, however, the lenticular component of the hologram may serve as an objective lens and can be an integral part of both angular and translational scan. This is given special attention in Section 2.8.

Geometrical scan design objectives relate primarily to the formation of a flat or a curved image field. The quest for a flat field is motivated by the practical desire to operate on a straight portion of the information medium. Thus "flat field" includes the (one-dimensional) straight-line scan traversing and focused on a surface having a coincident straight line, as on the surface of a drum and parallel to its cylindrical axis. Deviations from perfect flatness or perfect circularity are acceptable only to the extent that the information medium may be shaped to conform to those deviations, within acceptable tolerances on focal depth and linearity.

2.4. CHARACTERISTICS OF THE SCAN REGIONS

With these preparatory remarks, we now state some characteristics of the scan regions relating to the achievement of a flat field or a curved field with holographic or conventional scanners.

1. *When conducted by translation alone, objective scan yields an output which follows that of the translation.* This is true of a conventional quadratic phase function lens (see Section 4.4.3.1). The options are translation of the objective lens or translation of the information medium.

The important corollary is that translation of a collimated illuminating beam with respect to a *fixed* lenticular element will not alter the position of the focal point image: real or virtual. In Fig. 2.1 a collimated input beam may be represented by a reference point P_o situated at $-\infty$, whose finite movement is clearly irrelevant. Translation of a collimated beam is analogous to accessing a different portion of the same (larger) collimated beam. However, a practical means of accomplishing movement of P_i is movement of its conjugate object point P_o located a finite distance from the lenticular element—an option that may be implemented for focal-length change or transverse positioning of P_i.

2. *When angular scan is accomplished prior to a fixed objective lens, preobjective scan can provide a flat field.* The scanned flux is usually collimated and intercepted by an objective lens that converges the flux to a focal point and (serving as a flat-field lens) places the scanned locus (ideally) on a straight line normal to the principal scanned axis. When the scanner is holographic and its grating is uniform and linear (no optical power), a collimated reconstruction beam is diffracted to a collimated scanned output beam.

3. *When scan is conducted with an angular change following a fixed objective lens, postobjective scan will provide a generally arcuate (curved) locus about the rotating axis. Further, when the fixed objective lens converges its initial focal point on the rotating axis* (Bei 6), *postobjective scan will execute a perfectly circular locus about the axis.* This special case is described in Section 4.1.2. However, when the initial focal point is *not* coincident with the rotating axis, the scan locus forms a generally noncircular limaçon function (Bei 6).

4. *When scan is conducted with an angularly changing device to which an objective lens (segment) is mechanically coupled, operation is similar to that of postobjective scan. Further, when the input illumination is derived from a point on the rotating axis (radially symmetric), circularly symmetric scan will be executed about the axis.* This special case is described in Sections 4.1.2 and 4.1.4. This condition of the objective lens coupled to the angularly actuating mechanism, although rare in conventional scanning, is often operative in holographic scanning. Thus it will be given extensive attention in this book, notably in Section 2.8.

It is useful to divide the several hologram types into two categories: linear and lenticular. A *linear hologram* is effectively a linear diffraction grating on a plane surface, analogous (at a single wavelength) to a plane mirror or prism. A *lenticular hologram* is analogous to a curved mirror or lens. Since the lenticular hologram exhibits an average grating periodicity which creates an angular bias (e.g., off-center portion of a Fresnel lens), in its most general form it can be considered to include a linear component. This is analogous to the off-center portion of a conventional lens, which may be considered composed effectively of a prism (angular bias) in tandem with an axially symmetric lens (Bry 1). The significance of this categorization is that only the linear hologram can provide uniquely preobjective or postobjective scan functions. The lenticular hologram can exhibit characteristics of both objective and postobjective scans (per Section 2.8).

2.5. CONDITION OF OBJECTIVE LENS ON ROTATING SUBSTRATE

To introduce an important option in holographic scanning and the rare case with conventional components where the lenticular element rotates with the substrate, we construct in Fig. 2.2 three illustrative scanners. The first two are composed of conventional elements (mirrors and lenses) and the third represents a holographic equivalent. All three provide identical arcuate scan of point P along the circumference of a circular cylindrical information surface whose axis is coincident with the rotating axis. When the cylindrical medium is "unwrapped," the scans appear as straight lines. Either the cylindrical surface or

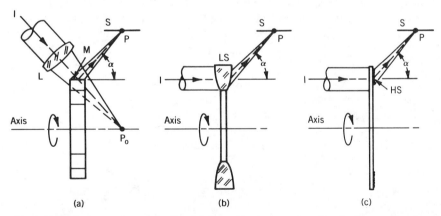

(a)   (b)   (c)

Figure 2.2. Analogous conventional and holographic devices executing circular scan locus on cylindrical information surface S concentric with axis. (a) Mirrored polygon overilluminated by beam I through fixed objective lens L. Original focal point intersects axis at P_o and is intercepted by mirror M to be reflected to point P. (b) Lens segments LS affixed to rotating disk. Illuminating beam I is refracted to form moving focal point P. (c) Hologram segments HS formed on rotating disk substrate. Illuminating beam I is diffracted through angle α and converged to form focal point P.

the scanner is customarily translated axially at uniform velocity with respect to the other to form an orthogonal raster. An alternate information surface is a plane oriented normal to the axis, upon which point P generates arcuate segments. For simplicity, all three are overilluminated (defined in Section 2.8.1), so that the full optical aperture establishes the scanning aperture. It may be appreciated at this level and will be shown in Section 2.8.3 that the scanned image dynamics are invarient with over- or- underillumination, even though the scan fulcrum moves from the axis (when overilluminated) to the aperture (when underilluminated).

Figure 2.2a is an example of postobjective scan, with the illuminating beam convergent at P_o on the rotating axis of a multifaceted (mirrored) polygon. As expressed in characteristic 3 of Section 2.4, this provides a perfectly circular arc of scanned point P, centered on the rotating axis, traversed with an angular velocity equal to that of the rotating substrate.

Uncommon as it appears, Fig. 2.2b executes precisely the same scan function. Here the collimated illuminating beam is equally coincident with the rotating axis (at infinity), providing a 1 : 1 scan angle-to-rotational angle relationship. The lens segments LS affixed to the substrate rotate with it, serving to redirect the incident collimated light radially and focus to a point P which executes the circular arc centered on the rotating axis. Figure 2.2a and b, unusual configurations, were selected to match the characteristics of the holographic scanner to follow.

Figure 2.2c shows a corresponding disk scanner. The collimated illuminating beam is similar to that shown in Fig. 2.2b, while the holographic sector HS serves to both tip the principal ray radially outward through the same angle α and provide lenticular (optical) power to converge the output flux to focal point P. This holographic element is thus seen to be equivalent to a marginal sector of a (zone or conventional) lens whose optical axis is parallel to the rotating axis and coincident with the focal point P. Figure 2.2b and c represent characteristic 4 of Section 2.4.

The examples above are all radially symmetric. When nonradially symmetric, the scan loci depart from these simple functions. An analogous construction can be made for nonradially symmetric scanners (e.g., Fig. 4.30).

2.6. RELATIONSHIP OF SUBSTRATE SURFACE TO AXIS

2.6.1. Stationary Orientation

The substrate that retains radial symmetry exhibits a surface which remains stationary during rotation about its axis. Referring now to the image area, a flat-field image surface is formed from either a plane or a cylinder (e.g., a photosensitive drum whose axis is parallel to the along-scan direction). (A cone-shaped image surface is also admissible.)

For an angularly scanned function to generate a straight-line image, it is necessary that the scanned arc reside in a plane such that the intersection of this plane with the image surface forms a straight line. To obtain a scanning function (from a radially symmetric system) that resides in a plane, we must derive an output whose principal rays propagate normal to the rotating axis. Otherwise, it generates a cone, as described in Section 4.1.4. Since, in general, a bowed scan into a lens generates a bowed image, and a planar scan through the meridional plane of the lens generates a straight-line image, it is most often arranged that the scanned arc reside in the meridional plane of the objective lens.

Whether transmissive or reflective, the ability of a hologram to diffract an output beam that propagates in a direction normal to the rotating axis is determined by the local orientation of the substrate with respect to the axis. Figure 2.3 shows a generic "kidney-shaped" rotationally symmetric substrate upon which is incident an illuminating beam I at a point having a surface normal $\bar{n}$. The output O can be derived normal to the z axis only when the surface presents a component $\bar{n}_\perp(z)$ which is, at the same time, normal to the z axis.

This rule has the equally clear corollary that a flat disk whose surface normal is parallel to the axis (most common mounting) can not diffract a beam (into free space) normal to the axis. This factor has, for a long time, frustrated the use of disk-type holographic scanners without the use of auxiliary optics (covered in Section 4.2.2.3) and complementary straight-line approximation in non-radially-symmetric systems (Section 4.2.2.5).

2.6.2. Dynamic Orientation

The task of the optical scanning device is to execute a well-controlled (usually cyclic) traversal of a radiation focal point through a fixed-image locus. The

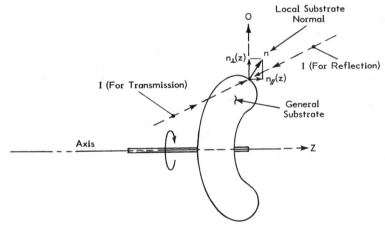

Figure 2.3. Section of general substrate symmetric about its z axis. Only when it presents a component of surface $n_\perp(z)$ can a diffracted output O be directed normal to the axis.

demands on this regularity are typically proportional to the number of elements of resolution conveyed by the scanning device over its full traversal: often specified as a small fraction (say, $\frac{1}{10}$) of the elemental spacing. Hence repeatability of the scan function could be required to be 1 part in 10,000 for 1000-element resolution, or 1 part in 100,000 for 10,000-element resolution, extending to a demand for cyclic accuracy to within 1 part in several hundred thousand. The bases of these dynamic factors are introduced here, with more detailed analysis provided in Chapter 5.

Since most image scans nominally execute straight lines along surfaces that are ultimately formed flat, rectangular coordinates conveniently designate the image area. We assign the x-direction to be the "along-scan" direction and the y direction to be the "cross-scan" direction. Although these are sometimes identified as "tangential" and "sagittal" directions, respectively, we retain the more direct "along-scan" and "cross-scan" descriptors throughout this book.

The array of scan lines arranged to cover an area generally forms a rectangular raster composed of nominally linear strokes in the along-scan direction which are equally spaced in the cross-scan direction. Because the along-scan nonuniformities are often continuous and slowly varying and those in the cross-scan direction appear in discontinuous groups, the perception of the latter error is most acute. Although such errors can be generated by a nonuniform medium transport in particular and by external forces in general, our principal concern is the error generated by scans that are not sufficiently congruent in the cross-scan direction.

The most prominent causative factor for cross-scan nonuniformity is a mechanical inconsistency termed "wobble," in which the plane defined by the scanner fails to sustain adequate stability. As a consequence, even though the illumination on the scanner may be stationary, the diffracted wave will fail to be sufficiently repetitive, generating unequal spacing among adjacent scan lines in the image plane. One of the most significant contributions to holographic scanning was the perception and analysis (Kra 3) that this wobble may be reduced many-fold under certain conditions of incident and diffracted ray angles with respect to the substrate of transmission holograms. It is accentuated by reflection holograms, much as by mirror reflection. This and other factors relating to mechanical/optical integrity are discussed further in Section 4.2.2.4 and Chapter 5.

2.7. CLASSIFICATION SUMMARY

Figure 2.4 provides a graphic representation of many of the options available in selecting and utilizing holographic scanners, as presented in this chapter and described in subsequent discussion. The holographic grating may be thin or thick, reflective or transmissive, independent of the balance of options. If a

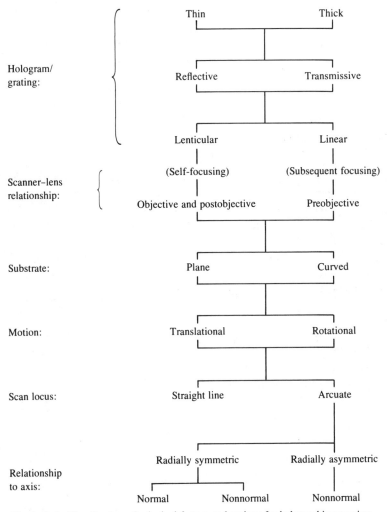

Figure 2.4. Classification of principal factors and options for holographic scanning.

lenticular grating, it is likely to be self-focusing and can serve objective scan or postobjective scan. If a linear grating, usually requiring subsequent focusing, it is classed as preobjective. All of the foregoing characteristics may be realized with plane or curved substrates operating in translational or rotational motion, providing either straight-line or arcuate scan loci. If arcuate, the illumination and scan locus may be radially symmetric with the axis, in which case the output flux could be normal to the axis. If radially asymmetric to the axis, it will probably also be nonnormal over a portion of its scan.

Variations, such as single and multiple outputs, over- and underillumination, and multidimensional scan will also be represented during related discussion, as will be the factors relating to achievement of linearity and resolution in the along-scan (tangential) and cross-scan (sagittal) directions.

2.8. SCANNED RESOLUTION

2.8.1. Introduction and Definitions

Although we are concerned primarily with the resolution provided by holographic scanners—and there are some special considerations that distinguish the analysis for such scanners—it will become apparent that the resolution of holographic scanners is governed by the same relationships that define the resolution of other optical scanners. This section reflects a dedication toward unification, for the resolution of holographic scanners has, at times, been expressed by responsible investigators as being composed of unique parameters (Pol 1) and/or complicated by extraneous factors (Ih 2) which we will subsequently discuss. The analogy between conventional and holographic lenticular elements is reinforced by an appreciation of their common focal-depth characteristics (Mei 2).

In formulating the resolution of holographic scanners and appearing to follow a nonholographic course, it will become evident that we are developing cognizance for the resolution of all optical scanners—holographic included. The expressions acquire extra significance for certain holographic scanners, for they are more likely to scan noncollimated waves (converging, for example) than are conventional ones. It will be shown that such noncollimated scans can be augmented in resolution—more or less, depending on the convergence or divergence of the scanned wave derived from either a conventional or a holographic device.

The resolution of an optical scanner is expressed (Bei 6, Zoo) by the number of elemental points that can be conveyed along a contiguous spatial path. Usually, the spatial path is nearly linear and traversed with uniform velocity, oriented along a relatively straight line, and distinguishing nominally uniformly spaced elemental points along its path. Although these points are analogous to the familiar descriptors "pixels" or "pels" (picture elements), we circumvent such identification, for pixels sometimes denote spatially digitized scan, where each pixel is uniform in intensity and/or color within its allotted area. Optical scan, on the other hand, when viewed along a single line is typically contiguous, except as it may be subjected to modulation (analog or quantized) depending on user orientation and application. Normally, therefore, the holographic scanner executes a uniform spatial function that can be divided into elements by modulation of its intensity. To avoid perturbation of the elemental point spread function (PSF) by the modulating (or sampling) process, we assume (when needed) that the contiguous scan is modulated in intensity with a series of (Dirac) pulses of infinitesimal width separated by a time t such that the spatial separation between imaged points is $w = vt$, where v is the uniform velocity of the scanned beam. We shall also define the "spot size" of the thus-established elemental function as δ and assume that $\delta = w$; that is, the width of the imaged spot is equal to the spacing from its neighbor. To limit further the variability of width definition, we consider the spot to be a generally rounded intensity main lobe having a width of magnitude δ measured at its 50% intensity points, sometimes represented as FWHM (full width at half maximum) of its PSF. Also, unless otherwise stated, we denote resolution N as being that derived from a diffraction-limited system. Aberration can affect the overall resolution significantly and must be anticipated and accommodated in any practical system design. Further, N is limited to the *active* portion of scan, that is, without the time or space commonly devoted to blanking or retrace intervals (sometimes called "latency"). If T is taken as the full scan period and τ as the blanked or retrace interval, the active portion for N resolution elements is $T - \tau$ and the duty cycle η_c is that remainder divided by the full period T. Thus

$$N = N_{\text{max}} \left(1 - \frac{\tau}{T} \right) = N_{\text{max}} \eta_c \qquad (1a)$$

where N_{max} is the ideal resolution that may be provided during unity duty cycle. As in other scanners, notably during overillumination of a multifaceted polygon (Bei 6), unity duty cycle can be achieved as well in holographic scanners. This is discussed further in this section and in Section 4.2.1.1. Designing Θ as that optical angle executed during the active portion of scan, then (for $\dot{\Theta} = d\theta/dt$ = constant)

$$\Theta = \Theta_{\text{max}} \left(1 - \frac{\tau}{T} \right) = \Theta_{\text{max}} \eta_c \qquad (1b)$$

To provide for underillumination, when a typically uniform facet (angular) velocity $\dot{D} \approx D/\tau = D_{\text{max}}/T$, where D_{max} is the full aperture, of which only the portion D is illuminated at any one time, we state more generally that

$$\Theta = \Theta_{\text{max}} (1 - \Delta) \qquad (1c)$$

in which $\Delta = \tau/T = D/D_{\text{max}}$.

We shall invoke two principles that permit more succinct expression of scanned resolution. One is that in considering angular scan, we avoid involvement with the spot size δ. While resolution may be defined as the number N of diffraction-limited δ-sized spots subtended by the format width W, it is also expressed by the number of diffraction-limited angles $\Delta\Theta$ contained within the full-scanned angular subtense Θ. That is, in a spatially linear system,

$$N = \frac{W}{\delta} = \frac{\Theta}{\Delta\Theta} \tag{2}$$

with $\Delta\Theta$ due to diffraction taken as

$$\sin \Delta\Theta = \frac{a\lambda}{D} \tag{3}$$

in which D is the illuminated aperture width and a is its shape factor (Bei 6), both measured in the direction of deflection. For the small diffraction angles propagating typically from $D \gg \lambda$, $\sin \Delta\Theta = \Delta\Theta$ and from Eqs. (2) and (3) we obtain directly

$$N = \frac{\Theta D}{a\lambda} \tag{4}$$

This is the classical expression for angularly scanned resolution.

To solidify the independence from subsequent optics, we recognize an important second principle of the Lagrange invariant (Lev), expressed in our nomenclature as

$$n\Theta D = n'\Theta'D' \tag{5}$$

where the primed terms are the refractive index, (paraxial) angular deviation, and aperture width, respectively, in the final image space, independent of intervening coaxial optical surfaces of revolution. For the common condition of $n = n'$ in air, the ΘD product and resolution N are conserved, invariant with centered optics following the deflector. Because of our subsequent discussion of augmentation, which relates to translational scan and deflectors displaced from the nutating axis, we restate Eq. (4) such that

$$N_\Theta = \frac{\Theta D_o}{a\lambda} \tag{6}$$

limiting N_Θ to resolution due to a radiated angular excursion from an aperture having a nutating fulcrum effectively in the plane of D_o. Θ is the active *deflected* excursion, not necessarily a mechanical one. To account for the optical radiation direction change, which may differ from a mechanical one which causes that change (such as that executed by D), we define a magnification

$$m = \frac{\Theta}{\Phi} \tag{7}$$

in which Φ is the corresponding mechanical change. As an example, when D_o is a (galvanometer) mirror having a nutating axis centered along its surface, $m = 2$ by reflection. In holographic scan, one may vary magnification as a function of the differing holographic exposure and reconstruction conditions, discussed in Sections 2.8.4 and 5.2.2.3.

We now distinguish between angular scan and translational scan by identifying

$$N_s = \frac{S}{\delta} \tag{8}$$

as the resolution due to *translation* over a distance S of a converging optical wavefront focused to a spot size δ (objective scan per Section 2.4). In general, therefore, a scanning system may accumulate resolution N due to two processes (which can occur simultaneously):

$$N = N_\Theta + N_s \tag{9}$$

due to an angular change N_Θ and to an equivalent translational one N_s. The diffraction-limited $\delta = aF\lambda$, in which a is the aperture shape factor and $F = f/D$ is the effective f number of the converging cone focused over the distance f from the aperture of width D.

All consideration thus far has addressed the "along-scan" resolution—that accumulated along the desired path of the nominal focal point. All relationships developed here relate only to that parameter. The operational scanning system, however, often forms a regular array of along-scan lines whose quadrature component is called "cross-scan." This array is called a *raster*. The resolution in the cross-scan direction is governed by additional factors which include line spacing and its uniformity, discussed in relation to anamorphic beam handling in Section 5.1. The spot size δ, which has thus far been measured along-scan, must exhibit a cross-scan component $\delta_\perp$. Since they are in quadrature, we invoke separable variables and consider them separately in Section 5.1. Nominally, $\delta_\perp \approx \delta$ to approach an isotropic spot in the image field. We shall at times refer to the cross-scan resolution $N_\perp$ and assume that $\delta_\perp \approx \delta \approx w \approx w_\perp$, in which $w_\perp$ is the line-to-line center spacing. In this book no other interrelationship is required or assumed between these parameters. A raster is illustrated, for example, in the "Fig. 2" portion of Fig. 4.27, where "horizontal" is the along-scan direction and "vertical" is the cross-scan direction.

Finally, in this introductory section, we firm the definitions of two frequently used phrases and concepts: overillumination and underillumination. They are also identified as overfilling and underfilling, respectively. In *overillumination*, the light flux encompasses the entire useful aperture (in the along-scan direction). This is usually implemented by illuminating at least two adjacent apertures such that (the active) one is always filled with light flux as it executes its path from one end of scan to the other. One may not only attain unity duty cycle, but allow resolution to be maximized for two reasons: (1) blanking or retrace may be zero; and (2) full aperture width D is operative throughout scan. The trade-off is, of course, the loss of illuminating flux beyond the aperture edges and attendent reduction in optical throughput efficiency. An alternative is to prescan (Urb 2) the light flux synchronously with the path of the scanning aperture such that it is filled with illuminating flux during its transit (seldom instituted, due to added complexity).

In *underillumination*, the light flux is directed to a fraction of the full optical aperture, such that this *illuminating subtense* delimits the useful portion D. Under this, a fairly prevalent condition, a finite and often substantive blanking or retrace interval results, thereby depleting the duty cycle (increasing instantaneous bandwidth for a given average data rate) and imposing use of a much larger active deflecting component. In Section 5.1.2.1 we describe an anamorphic beam-handling method for increased duty cycle which is unique to holographic scanning. Underillumination gains primarily in throughput illuminating efficiency because almost all the flux arriving at the aperture may be utilized.

2.8.2. The Displaced Deflector and Augmented Resolution

The angular deflector displaced from its nutating axis is the dominant method for holographic scan. Its prototype is the familiar mirrored polygon with mirrors either at an angle to the axis (pyramidal or conic polygon) or parallel to the axis (prismatic polygon). Depending on the characteristics of illumination (Bei 6), the principal consequence of these varieties, for our current interest, is the variation of the angular magnification $1 \leq m \leq 2$. The same range of variation is

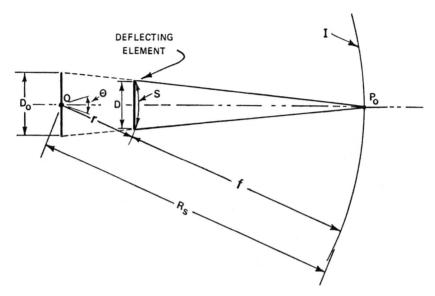

Figure 2.5. Deflecting element of width D displaced from rotating axis O by distance r. Output flux is converged to focus on surface I over distance f.

experienced by holographic scanners, although the reasons are somewhat different (discussed subsequently).

We start with the condition $m = 1$. Figure 2.5 illustrates the plane projected normal to the scanned surface, in which the output flux converges to point P_o as launched over a distance f from an aperture D. This deflecting aperture, which could be the hologram, is displaced from the fulcrum o by the distance r, such that $r + f = R_s$. Assuming overillumination, as exemplified in all cases illustrated in Fig. 2.2, analysis is simplified while providing the same results as for underillumination (see Section 2.8.3). By similar triangles,

$$\frac{D_o}{D} = \frac{R_s}{f} \tag{10}$$

and

$$D_o = D\left(1 + \frac{r}{f}\right) \tag{10a}$$

Substituting Eq. (10a) into Eq. (6) yields

$$N_o = \frac{\Theta D}{a\lambda}\left(1 + \frac{r}{f}\right) \tag{11}$$

$$= \frac{\Theta D}{a\lambda} + \frac{\Theta D r}{a\lambda f} \tag{11a}$$

There now appears an augmentation term $\Theta D r / a\lambda f$ which elicits interpretation. From Fig. 2.5 the radian arc length $S = r\Theta$, whence Eq. (11a) reduces to

$$N = \frac{\Theta D}{a\lambda} + S\frac{D}{a\lambda f} \tag{12}$$

$$= \frac{\Theta D}{a\lambda} + \frac{S}{\delta} \tag{12a}$$

$$= N_\Theta + N_s \tag{12b}$$

in which δ is the diffraction-limited spot size. We have constituted the hypothesized Eq. (9), which indicates that the total resolution is the sum of that due to angular change N_Θ and that due to equivalent translational scan N_s. When the deflector is displaced from the axis, it executes a translation $S_t \approx r\Theta$ to form resolution N_s, while at the same time executing the angular change Θ to form resolution N_Θ.

The basis for this augmentation is further appreciated from Fig. 2.5 when we realize that the deflecting aperture D which is displaced by r from the axis may be equally represented by the deflecting aperture D_o, whose fulcrum is *at* the axis. And [by Eq. (10a)] D_o is greater than D by precisely the same factor as is the resolution $(1 + r/f)$ since resolution is proportional to the width of the deflecting aperture.

Augmentation reduces to zero when $r = 0$ (deflector fulcrum on axis) and/ or when $f = \infty$ (output beam collimated). Further, augmentation is positive (resolution additive) when f is finite and positive (beam converging to real image point). When f is negative (beam *diverging* from virtual image point), we conclude the rather surprising result that augmentation is *negative* and resolution is *decreased*.

This case of a diverging beam is not to be interpreted as expanding due to diffraction alone. That is, if the beam propagates collimated from the hologram but diverges due to diffraction (particularly from a "small" aperture), the focal length *at the aperture* is effectively $f = \infty$ and augmentation remains zero. The focal length is finite (positive or negative) only when so purposefully directed, as from an aperture having lenticular power which is illuminated by a collimated beam. In holographic scanning systems, the aperture is more likely to exhibit lenticular power than in any other scanning mechanism. Although the existence of such lenticular elements does not necessarily require that the output wavefront be diverging or converging (for incident illumination can be such as to provide a collimated output), the prevalent utility of lenticular holographic elements in scanning applications is to form a converging wavefront into a real focal point at a remote image surface. Thus positive augmentation exists frequently in holographic scanning.

Negative augmentation develops when the output beam is diverging. Or it could be converging and drawn from a segment of the scanner that is radially opposite its propagation direction, whereupon r is negative and the augmenting term becomes $(1 - r/f)$. Such a special case is described in Section 4.2.2.2.1, where we discuss the concave Holofacet scanner.

2.8.3. Augmented Scan of the Underilluminated Aperture

To include explicitly the equivalent augmentation of underilluminated apertures, consider Fig. 2.6, which represents a (holographic) positive lenticular element in translation displaced a distance $s/2$ along the x axis. As delimited by the smaller aperture d, the flux is converged over the (same) distance f to a (larger) spot size

$$\delta_d = a(f/d)\lambda = aF_d\lambda \tag{13}$$

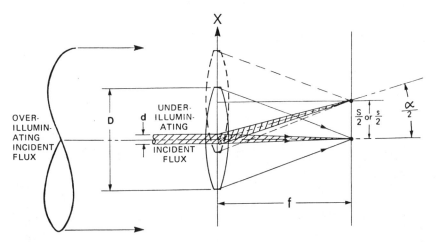

Figure 2.6. Positive lenticular element executing translation scan. Overilluminated, aperture D delimits the flux tube; underilluminated, it accepts the smaller flux tube d.

where F_d is the f-number enlarged by the restricted aperture d. The resolution, although reduced, remains defined as

$$N_d = \frac{s}{\delta_d} \tag{14}$$

in which $s = D - d$. The output scan function now appears *angularly displaced* through $\alpha = s/f$. Substituting $s = \alpha f$ and Eq. (13) into Eq. (14) yields

$$N_d = \frac{\alpha d}{a\lambda} \tag{15}$$

which corresponds to the basic relationship of Eq. (4) for an angular scan.

If the same lenticular element is arranged to execute an angular displacement per Fig. 2.7, extended from the fulcrum by the radial distance r, the image point from d effectively executes an additional arced distance s. The angular component α above now augments the angular resolution attributed to Θ alone, forming

$$N_{d\Theta} = \frac{(\Theta + \alpha)\, d}{a\lambda} \tag{16}$$

$$= N_{\Theta d} + N_d \tag{16a}$$

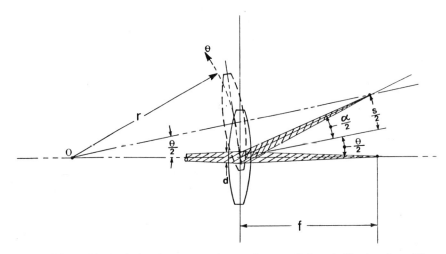

Figure 2.7. Positive lenticular element executing angular scan. Only underilluminated condition shown, for simplicity.

Since

$$s = \alpha f = r\Theta \tag{17}$$

then

$$\alpha = \Theta(r/f) \tag{17a}$$

which when substituted into Eq. (16) yields

$$N_{d\Theta} = \frac{\Theta d}{a\lambda}\left(1 + \frac{r}{f}\right) \tag{18}$$

We have reconstituted the augmented relationship of Eq. (11), differing only in the d versus the D, which denote under- versus overillumination, respectively. Accordingly, we express the more general resolution equation for an angularly scanned aperture D (in unity magnification),

$$N = \frac{\Theta D}{a\lambda}\left(1 + \frac{r}{f}\right) \tag{19}$$

in which the D denotes the aperture size, with no restriction on its illumination filling.

In these and subsequent expressions of the angle Θ, an unambiguous definition is provided when it is evaluated as though the aperture were overilluminated. This is a test condition for Θ, not a condition for operation. In Fig. 2.7, for example, $\Theta/2$ is the angular change of the principal ray, and uniquely excludes the augmented angular change $\alpha/2$.

2.8.4. Generalized Resolution, Including Scan Magnification

Thus far we have restricted resolution consideration to the case of $\Theta = \Phi$; that is (when tested as overilluminated), the output scan angle equals the mechanical one. This is exemplified by many holographic scanners in which the mechanical and optical radii are common. To include those in which the angles are not equal, where the hologram may be illuminated from a point differing from its effective construction source location, or when the output scan plane differs from the plane projected normal to the rotating axis, the equations need to be extended (Bei 12).

Two cases are illustrated, in Fig. 2.8a with unity magnification and in Fig. 2.8b with magnification $m = \Theta/\Phi > 1$. The deflecting element is shown operating in transmission, as for a hologram on a drum. Any deflecting element

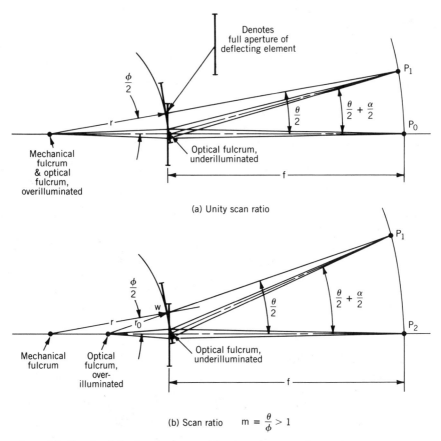

Figure 2.8. Rotating deflecting element providing (*a*) unity scan ratio and (*b*) scan ratio $m > 1$, shown underilluminated. When overilluminated, full aperture delimits output beam, of which only principal ray shown, for simplicity.

(such as a mirrored polygon) is applicable when considering the effect on the output flux. Also, if derived from a device whose output is not radially directed but which exhibits a component in the axial direction (such as the conic scan from a disk-like device), the representation remains valid as the projection of scan on a plane normal to the rotating axis (in the plane of the paper of Fig. 2.8) if the output is accepted as arcuate. Other conditions for the establishment of parameters, including those for unbowed scans from disk substrates, are provided at the close of this section.

Taking w as the common displacement at the intersection of r and r_o in Fig.

2.8b, then $\sin \Phi_{1/2} = w/r$ and $\sin \Theta_{1/2} = w/r_o$, where the subscript $\frac{1}{2}$ denotes the half-angle from the center of deflection. Then

$$r_o = r \frac{\sin \Phi_{1/2}}{\sin \Theta_{1/2}} = \frac{r}{m'} \qquad (20)$$

where

$$m' = \frac{\sin \Theta_{1/2}}{\sin \Phi_{1/2}} \qquad (20a)$$

The new scan radius r_o of Eq. (20) now replaces the original r in Eq. (19), yielding

$$N = \frac{\Theta D}{a\lambda} \left(1 + \frac{r}{m'f}\right) \qquad (21)$$

For parametric analysis, series expansion of Eq. (20a) to the second significant term yields

$$m' = m \left(\frac{1 - \Theta_{1/2}^2/6}{1 - \Phi_{1/2}^2/6}\right) \qquad (22)$$

allowing Eq. (21) to be approximated by

$$N = \frac{\Theta D}{a\lambda} \left[1 + \frac{r}{mf} \left(\frac{1 - \Phi_{1/2}^2/6}{1 - \Theta_{1/2}^2/6}\right)\right] \qquad (23)$$

To determine the significance of the term in parentheses in Eq. (23), when it modifies r/mf with a factor that imposes less than a fractional error ϵ, we set

$$\frac{1 - \Theta_{1/2}^2/6m^2}{1 - \Theta_{1/2}^2/6} < 1 + \epsilon \qquad (24)$$

and solve for the bound on $\Theta_{1/2}$, yielding

$$\Theta_{1/2} < \left(\frac{6\epsilon}{1 + \epsilon - 1/m^2}\right)^{1/2} \qquad (25)$$

Evaluating, one finds that substantive scan angles of $\Theta = 70°$ and $96°$ are allowed for $m = 3$ and 2, respectively, before imposing $\epsilon = 0.1$ influence on the fraction r/mf. When $m \rightarrow 1$ (i.e., $\Theta \rightarrow \Phi$), Θ may be extremely large, even for very small ϵ.

The influence of a finite ϵ on the resolution N derives from Eqs. (23) and (24),

$$N = \frac{\Theta D}{a\lambda} \left[1 + \frac{r}{mf} (1 + \epsilon) \right] \qquad (26)$$

whence, expressing the bracketed term as

$$\left(1 + \frac{r}{mf} \right) + \left(\frac{r}{mf} \epsilon \right) \qquad (27)$$

ϵ is seen to be diluted further by the factor $r/mf < 1$ such that the second term $\ll 1$ while the first term is almost always >1. For example, when $r/mf = 0.1$ while $\epsilon = 0.1$, the second term is less than $\frac{1}{100}$ of the first, imposing less than 1% variation on the resolution N. The influence of m must be considered primarily when greater than 2, for only then does it approach constraint upon practical scan angles. And the larger m is, the smaller is the second term (other factors being equal), making the influence of ϵ even less significant. This leads to a simplified expression for general scanned resolution, valid over a very wide operating range:

$$N = \frac{\Theta D}{a\lambda} \left(1 + \frac{r}{mf} \right) \qquad (28a)$$

or

$$N = \frac{\Phi D}{a\lambda} \left(m + \frac{r}{f} \right) \qquad (28b)$$

Equation (28b) expressed the magnification m and the augmenting term r/f as more independent modifiers of the resolution N, a form not immediately available from Eq. (21).

We can express Θ or Φ in terms of Θ_{max} or Φ_{max} by including the duty cycle $n_c = 1 - \Delta$, in which $\Delta = \tau/T = D/D_{max}$. Θ is represented by Eq. (1c), while the equivalent for Φ is $\Phi_{max} n_c = \Phi_{max}(1 - \Delta)$. The duty cycle is not a function of m, since $\eta_c = \Theta/\Theta_{max} = m\Phi/m\Phi_{max}$. In conventional scanning, for example, a prismatic polygon operating at $m = 2$ and a pyramidal polygon operating at $m = 1$ both provide the same duty cycle for a given D/D_{max}.

To establish the parameters for magnification, two techniques of holographic scan magnification apply. In one, the reconstruction source radius r_o is selected shorter than that used in holographic construction, as represented directly in Fig. 2.8b. While scan angle is magnified by the ratio $m = r/r_o$, if the grating is nonlinear, the aberrations resulting from misalignment of reference and reconstruction waves need attention. Complex compensation techniques were developed in an early investigation (Lee 1), raising questions regarding the utility of this form of magnification, when stigmatic reconstruction (at the same wavelength) is achieved over a larger angle simply by reducing the number of holographic facets in a properly aligned system (Bei 8). The other method that provides magnification m tends toward Bragg operation for unbowing the scan line, illuminating at angle Θ_i and diffracting the output at angle Θ_o with respect to the hologram normal. When $\Theta_i = \Theta_o = 45°$, a magnification $m = \sqrt{2}$ results near the center of scan, determined by $m = \csc \Theta_o = \lambda/d$ (Kra 7; Appendix 3). The corresponding development (Ant; Appendix 2) is described in Section 4.2.2.5.1.

From Eq. (40) in that section, when $\Theta_i = \Theta_o = \alpha$ and scan angle $\Theta \rightarrow 0$, then $m = 2 \sin \Theta_o$ near the central portion of scan. When $\Theta_i = 0$ (radially symmetric), the output beam rotates directly with the substrate, and the scan angle is foreshortened such that $m = \sin \Theta_o$. Heuristically, then, a general expression for m at the central portion of the scan interval of a transmissive disk configuration is

$$m = \sin \Theta_i + \sin \Theta_o = \lambda/d \qquad (29)$$

which is the grating equation. This is developed more rigorously in Section 5.2.2.3.

2.8.5. Summary

Angularly scanned resolution (measured by the number of contiguous diffraction-limited imaged elements) is established only by action at the deflecting aperture on the output wave. It is independent of aperture over- or underillumination (for a given aperture subtense), independent of imaged spot size, and independent of subsequent passive optics (stigmatic and nonvignetting). Section 4.2.2.3.2 exemplifies the simplifications that can be provided for apparent analytically awkward and complex scanner configurations.

3

Holographic Basics and Techniques

3.1. WAVEFRONT AND RAY DIRECTION CONVENTIONS

In discussing the propagation of (light) radiation and its consequences, the concept of wavefronts and ray directions will be employed routinely. The equiphase surfaces that represent the (coherent) wavefronts are at all points (in homogeneous isotropic media) normal to the local ray directions and spaced apart along the ray or propagation direction by a distance equal to the wavelength of the quasi-monochromatic radiation. Unless otherwise indicated, the radiation is considered sufficiently monochromatic to register well-defined diffraction effects, and two or more are sufficiently coherent to register strong interference effects. The sign convention is provided at the close of this section.

3.2. THE GRATING EQUATION

The linear summation of the amplitudes of two mutually coherent waves forms an interfering standing-wave pattern with maxima and minima appearing at their common nodes and antinodes. Consider the incidence of two plane-parallel waves on a photosensitive surface—one arriving normal and the other at an angle Θ_o to the plane surface, as represented in Fig. 3.1. Due to time averaging, if either wave were incident alone, each would impart a gross effect on the surface, representing the averaged consequence of the exposure energy (intensity integrated over the exposure time)—as in the "fogging" of a photographic plate. When two intersect, however, a standing-wave pattern appears as a moiré (Shu) at the nodes of the two plane periodic waves. The pattern is also time averaged such that when traversing the plane (photosensitized) surface, the standing-wave intensity distribution imparts a periodic structure whose periodicity d (see the inset) is independent of the position of the surface—as long as

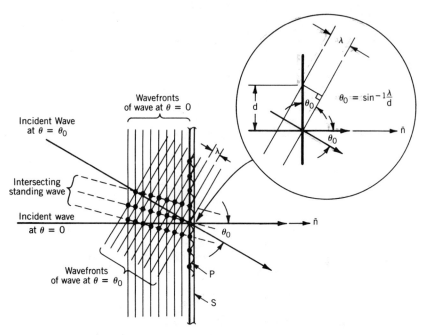

Figure 3.1. Mutually coherent collimated waves of wavelength λ incident on plane surface S having normal $\bar{n}$. One wave is normal and the other is at angle Θ_o, developing interference pattern P of period d on surface S.

the incident angles are sustained at wavelength λ. Only the phase orientation of the periodic structure will be determined by the position of the surface.

From this elementary model, we express a principal component of the grating equation,

$$\sin \Theta_o = \frac{\lambda}{d} = \nu\lambda \qquad (1)$$

where ν is the grating spatial frequency. This structure acts as a diffracting element, angularly displacing a wave incident at zero degrees to one at an angle Θ_o. We highlight the important observation that a linear displacement of this linear grating in any direction parallel to its surface alters only the phase of the grating structure relative to coordinates fixed to the illuminating wave, imparting a (typically imperceptible) optical frequency shift (Ste); it will not alter the diffracted angle Θ_o. If the grating subtense is uniform and large compared to the subtense of a narrow illuminating beam, such translation will register no measurable difference in the diffracted beam. If, however, the grating subtense is small compared to that of a broader illuminating wave, such translation will register as a parallel movement of the diffracted beam, displaced along with the grating. Finally, underscoring the importance of this simple model, if the grating is rotated about an axis parallel to the illuminating beam, the diffracted beam will rotate with the grating, maintaining the same angle Θ_o with respect to the axis, in the manner illustrated in Fig. 2.2c.

Thus far we have described only one diffracted component emanating from the grating surface. As expressed in the generation of the grating, Fig. 3.1 illustrates a sinusoidal interference pattern and is assumed recorded with a corresponding sinusoidal transmittance and/or phase variation. With no preference for the direction in which the diffracted beam will propagate, it splits into two equal components (for the thin sinusoidal grating): a fraction at $+\Theta_o$ and a complementary and equal fraction at $-\Theta_o$ with respect to the surface normal $\mathbf{n}$. This provides a simple extension to Eq. (1):

$$\sin \Theta_o = \pm \frac{\lambda}{d} \qquad (2)$$

and all conditions described above apply to both diffracted components, appearing as mirror images of each other about the grating normal $\mathbf{n}$.

When the grating is nonsinusoidal in amplitude and/or phase, constructive interference occurs at integral multiples of λ. This may be represented graphically as an extension of that in Fig. 3.1, with the multiple length $m\lambda$ forming another simple extension,

$$\sin \Theta_o = \pm m \frac{\lambda}{d} \qquad (3)$$

where m is a positive integer and the $+$ and $-$ signs denote the orders that emanate in pairs from nonsinusoidal thin gratings. Particularly enlightening illustrations and a discussion of this process are provided by Shulman (Shu).

Equation (3) is to include m, even in the sinusoidal case, to accommodate the zero-order diffraction. This output wave (when $m = 0$) is direct transmission of a fraction of the incident wave (unmodified in angle), manifest to make $\sin \Theta_o = 0$ (for the incident wave in this model is also at zero angle to the surface normal).

Allowing for arbitrary orientation of an incident wavefront at Θ_i with respect to the surface normal, and extending the modeling such that $d \sin \Theta_i + d \sin \Theta_o$

must now be equal to $m\lambda$, one derives the general grating equation in scalar form*:

$$\sin \Theta_i \pm \sin \Theta_o = m \frac{\lambda}{d} \qquad (4)$$

where, as is customarily interpreted, m is taken as a positive or a negative integer. The $\pm$ sign denotes directional convention for the angles (discussed in the next section): whether in the same or opposite sense from the surface normal and whether for reflection or for transmission gratings. For transmission gratings, the upper sign applies; for reflection, the lower applies.

The Bragg equation (Bra 1) is derivable directly. This relationship applies in the important case of equal input and output angles, discussed further in Section 4.2.2.4 for disk scanners operating in this Bragg regime. In transmission, with Θ_o clockwise in Fig. 3.2b, when $\Theta_i = \Theta_o = \Theta_b$ (the Bragg angle), we have from Eq. (4), $2 \sin \Theta_b = \lambda/d$, or

$$\sin \Theta_b = \frac{1}{2} \frac{\lambda}{d} \qquad (5)$$

3.3. SIGN CONVENTION

Sign conventions vary (And, Job). We shall maintain one that is prevalent in this technology. Θ_i is taken as positive, observing its orientation with respect to the surface normal. Then (Fig. 3.2a) the polarity of Θ_o is positive when directed in the opposite sense from the normal (i.e., with Θ_i counterclockwise from **n**); thus Θ_o is taken positive when clockwise from **n**. The sign convention for reflection or transmission can be established by solving for the zero-order output. Thus, in reflection, since it is required that $\Theta_o = \Theta_i$ (at $m = 0$) to satisfy Eq. (4), $\Theta_o - \Theta_i = 0$ and the lower sign applies. In transmission (Fig. 3.2b), where both angles are in the same sense with respect to **n**, $\Theta_o = -\Theta_i$ for the undeviated (zero-order) transmission, whence $\Theta_o + \Theta_i = 0$ and the upper sign applies.

3.4. THE ZONE LENS

A fundamental analog to the point-source holographic lens is the Fresnel zone lens (Nis, Suh, Swe, Col). As the grating equation described above forms an

*For this expression in vector form, see Ref. Clo 1.

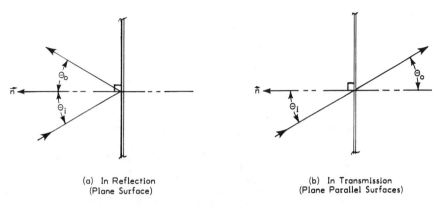

(a) In Reflection
(Plane Surface)

(b) In Transmission
(Plane Parallel Surfaces)

Figure 3.2. Establishment of sign convention from zero-order propagation. Θ_i always positive; Θ_o positive when directed in opposite sense from normal **n**.

important basis for the subsequent content of this book, so the zone lens solidifies our understanding of the nature of the holographic elements that provide optical power while executing scan. There are some formal distinctions between interpretations of zone lenses and their variations, some of which fall into the class of computer-generated holograms (also discussed later), all motivating a clear representation of this classic component.

The zone lens comprises an aperiodic array of (circular or linear) phase-altering elements, usually in the form of opaque and transparent regions called Fresnel zones. Examples of circular and linear (one-dimensional) zone lenses having alternating white and black spacings are illustrated in Fig. 3.3. While the zones illustrated in the figure exhibit approximately equal adjacent bands of opacity and transparency, it should be clear that the first-order imaging characteristics depend on their relative periodicity rather than on the detailed transmission characteristics within individual zonal convolutions. Radiometric efficiency and higher-order harmonic content are, however, influenced directly by their detailed transmission functions (Nis).

3.4.1. Infinity-Conjugate Zone Lens

To develop the zone lens relationships, we follow a procedure similar to that for the grating equation, imposing constructive interference at the nodes of the wavefronts (at least one being spherical) incident on a plane surface. For the analog of the in-line Gabor or Fresnel hologram (collimated background) or the lens operating with collimated incident light, we construct per Fig. 3.4 a collimated wave and a converging (spherical) wave arriving from the left, both symmetric with the axis and incident on the plane photosensitized medium M

(a)

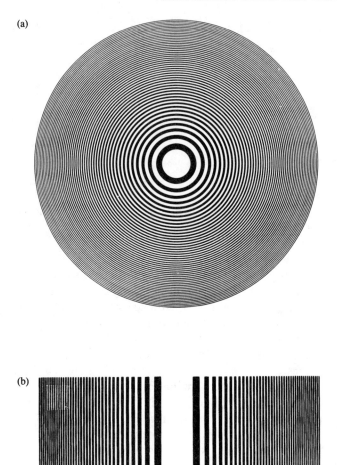

(b)

Figure 3.3. Photographs of Fresnel zone plates. (*a*) Circular zone plate (clear center), (*b*) Linear zone plate (clear center).

mounted normal to the axis. It may be recognized that in this two-dimensional medium, there is no distinction between the (thin developed) interference patterns whether the waves arrive from the same or from opposite directions.

With the collimated wavefront coincident with the plane surface M, for constructive interference, an integral number of waves $n\lambda$ must be added to f at each r_n. Since $f + n\lambda$ is the hypotenuse of the right triangle $r_n o\, f$, then

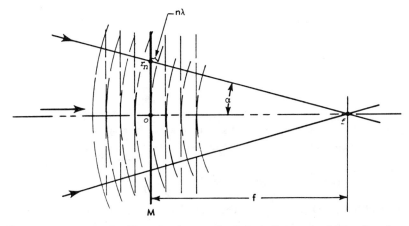

Figure 3.4. Wavefronts of collimated and converging (spherical) waves in vicinity of medium M for construction of zone lens.

$r_n^2 + f^2 = (f + n\lambda)^2$, yielding directly the quantized radii for wave reinforcement,

$$r_n^2 = 2nf\lambda + (n\lambda)^2 \tag{1}$$

When

$$(n\lambda)^2 \ll 2nf\lambda \tag{2a}$$

then

$$f \gg n\lambda/2 \tag{2b}$$

and we yield the *small-angle approximation* for the (full-period) radii of a zone lens

$$r_n^2 = 2nf\lambda \tag{3}$$

For the half-period zones m, convenient for alternating equal (area) clear and opaque circular rings (per period), then $n = m/2$ and $r_m^2 = mf\lambda$. For subsequent consideration of the phase functions, it is more useful to remain with the full-period zones.

To show and evaluate the small-angle approximation, combining Eqs. (2b) and (3) requires that

$$r_n \gg n\lambda \tag{4}$$

Referring to Fig. 3.4, the radius r_n will be much larger than the added wave group $n\lambda$ when the angle α is sufficiently small. To evaluate, with $\csc \alpha - \cot \alpha = n\lambda/r_n$, we assign $r_n > 10n\lambda$ to satisfy Eq. (4), whence $\csc \alpha - \cot \alpha < 0.1$, yielding (from series approximation)

$$\alpha < 0.2 = 11.5° \tag{5}$$

When the subtended half-angle is less than $11.5°$, Eq. (3) provides a reasonable representation of the zone lens. An awkward notation and use of Eq. (3) has appeared (You) to represent a geometric zone lens (GZP), while Eq. (1) represents validly an interferometric zone lens (IZP). The IZP, when properly constructed, will automatically form zones according to Eq. (1). It is equally expected that a zone plate constructed geometrically will also form zones according to Eq. (1). Equation (3) is normally applied only when the small-angle condition is satisfied, whether computed or interferometric. This gains importance for the more current technique of photoreducing a computer-plotted zone lens compared to the earlier master drawing of opaque and clear concentric circles.

A misleading variant appears in a holographic scanning analysis (Ono 1) interpreting that an IZP is aberrating in its marginal region because it fails to meet the requirement expressed by Eq. (3), identified as that for a GZP. This elicited clarification (Bei 11). It is often important to maintain integrity per Eq. (1) during holographic scanning, when such a diffractive lens element is displaced significantly with respect to a smaller illuminating beam. To reduce ambiguity of nomenclature in this age of computational geometric structure, the designation CZP is offered to denote the computed form, and LZP to represent the limited zone plate constructed according to Eq. (3).

The phase function $\Phi(x, y)$ is determined by setting n as an integral number of waves, $n = \Phi(x, y)/2\pi$, and substituting into the original relation $r^2 + f^2 = (f + n\lambda)^2$, yielding

$$\Phi(x, y) = \frac{2\pi}{\lambda} \left[(r^2 + f^2)^{1/2} - f \right] \tag{6}$$

For the small-angle approximation of Eq. (3), the substitution of $n = \Phi/2\pi$ yields

$$\Phi(x, y) = \pi r^2 / f\lambda \tag{7}$$

remaining reasonably accurate only when $\alpha < 0.2$ per Eq. (5).

3.4.2. Finite-Conjugate Zone Lens

The point image finite-conjugate lens derives as an extension to the in-line Fresnel or Gabor type expressed above. It exhibits properties similar to those of a conventional duo-convex lens positioned in-line within its conjugate image points. The familiar thin-lens formula applies,

$$1/f = 1/f_o + 1/f_i \tag{6}$$

where f_o and f_i are the distances to the object and image points, respectively, on either side of the principal planes. The equivalent construction for the conjugate-point zone lens appears in Fig. 3.5.

Constructive interference requires that the sum of the two hypotenuses (which meet at r_n) be equal to the sum of the two axial focal lengths plus the periodically reinforced wavelengths $n\lambda$. Thus

$$\left(r_n^2 + f_o^2\right)^{1/2} + \left(r_n^2 + f_i^2\right)^{1/2} = f_o + f_i + n\lambda \tag{7}$$

As for the infinity-conjugate zone lens, setting $n = \Phi'(x, y)/2\pi$ and substi-

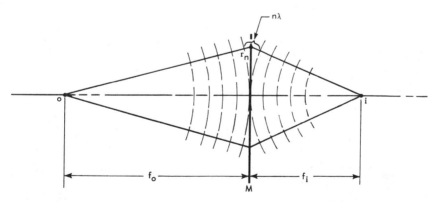

Figure 3.5. Wavefronts of two interfering spherical waves in vicinity of medium M for construction of finite-conjugate zone lens.

tuting into the relation above, the finite-conjugate phase function is determined directly as

$$\Phi_c(x, y) = \frac{2\pi}{\lambda} \left\{ \left[(r_n^2 + f_o^2)^{1/2} - f_o \right] + \left[(r_n^2 + f_i^2)^{1/2} - f_i \right] \right\} \quad (8)$$

When the object and image points are symmetric about the lens, this reduces to (for $f_o = f_i = 2f$)

$$\Phi_s(x, y) = \frac{4\pi}{\lambda} \left[(r_n^2 + 4f^2)^{1/2} - 2f \right] \quad (9)$$

Expressing this symmetric zone lens in terms of the zone radii, one finds that

$$r_s^2 = 2nf\lambda + \left(\frac{n\lambda}{2} \right)^2 \quad (10)$$

When compared to Eq. (1), we see identity, except for the division by 2 in the small "correction term," making it even smaller. This motivates some investigation, which we undertake in the following section.

3.4.3. Generalized Zone Lens

An important generalization to the zone lens has been expressed recently (Ono 2) and expanded (Ono 3), both works yielding significant performance from rotation of segments of these adjusted lenses. The works are discussed in Sections 4.2.2.6.2 and 4.2.2.6.3.

In Eq. (10) appeared a factor 2, while in Eq. (1), the corresponding factor would be 1. Equation (10) is formed by the interference of two spherical waves, while Eq. (1) required one spherical wave. Let us express that factor as a variable, stating

$$r_N^2 = 2nf\lambda + \left(\frac{n\lambda}{N} \right)^2 \quad (11)$$

where N is the new parameter and n remains as the fringe number, and consider some of its consequences.

If we extend $N \to \infty$, the second term drops out to form Eq. (3) for the limited zone plate (LZP). Equation (11) now represents three known "classical" configurations of zone lenses: formed in-line with one spherical wave (N

= 1), two spherical waves ($N = 2$), and an infinite number of spherical waves ($N = \infty$). The objective for the variation in N is the quest for adjustment of the fringe spacing of the zone lens in a controlled manner, to accommodate special focal characteristics. This hypothetical process, which might at first appear impossible, was actually instituted interferometrically with successive holographing of holographic zone lenses: one case having $N = 3$ (Ono 2) and another having $N = 4$ (Ono 3). In the first case, an $N = 2$ hologram is made with symmetric object and image points, as introduced for Eq. (9) having fringes according to Eq. (10). It is copied holographically with plane waves to select the desired diffracted first-order component, and that copy is then used to provide the $N = 2$ wavefront for a new symmetric hologram, interfering with an additional spherical wave, forming $N = 3$. In the second case, seeking $N = 4$, an $N = 2$ hologram is made initially, as described above. It is copied holographically with plane waves as above, but this time, selecting the second-order diffractive component. That second hologram, which represents $N = 4$, is then used to provide the object wave for the final hologram, interfering with a plane reference wave (no additional spherical wave)—forming the operational $N = 4$ hologram. Details on illumination directions, orientations for order selection, materials, and processing procedures toward replica fabrication appear in the works cited. Following a logical progression, one can see the tandem-exposure formation of almost any finite N-tuple hologram.

Effectively, we are superposing coherently, in the plane of the new hologram, the phase variations from a succession of single spherical waves, each having phase variations represented by Eq. (6). This leads to the generalized phase variation,

$$\Phi_N = \frac{2\pi}{\lambda} \sum_{k=1}^{N} \left[(r^2 + f_k^2)^{1/2} - f_k \right] \tag{12}$$

in which the individual focal lengths are f_k and the composite focal length is

$$\frac{1}{f} = \sum_{k=1}^{N} \frac{1}{f_k} \tag{13}$$

This new interferometrically formed zone lens creates a problem in nomenclature not yet addressed elsewhere. Either we retain the IZP descriptor to represent uniquely the case of $N = 1$ and call the rest something else, or we generalize IZP to identify its number N of equivalent spherical waves. We offer the IZP$_N$ approach, where N is assumed 1 when represented by Eq. (6) and includes the subscript to identify its formulation when $N \neq 1$. This is similar to the practice in diffraction processes of implicit assumption of first order and explicit identification of the higher orders.

To expand on the general form, we can also express a new grating vector, defined as (Col)

$$\mathbf{K} = 2\pi/d \qquad (14)$$

where d is the grating spacing taken in the direction normal to the maximum grating change. For circular zones, this is the radial direction, for which we know from Eq. (1) in Section 3.2 that

$$\frac{\lambda}{d(r)} = \sin\Theta_o = \frac{r}{(r^2 + f^2)^{1/2}} = \nu(r)\lambda \qquad (15)$$

whence, for the IZP,

$$\mathbf{K}_I = \frac{2\pi}{\lambda}\frac{r}{(r^2 + f^2)^{1/2}} \qquad (16)$$

and for the LZP,

$$\mathbf{K}_L = \frac{2\pi}{\lambda}\frac{r}{f} \qquad (17)$$

where the factor multiplying the wave number $2\pi/\lambda$ is the sine of the diffracted output ray with respect to the axis. For the IZP_N, the grating vector becomes

$$\mathbf{K}_N = \frac{2\pi}{\lambda}\sum_{k=1}^{N}\frac{r}{(r^2 + f_k^2)^{1/2}} \qquad (18)$$

3.5. HOLOGRAM FORMATION AND WAVEFRONT RECONSTRUCTION

In this section we introduce some of the fundamental properties of light-wave interaction with a photosensitive medium, and the consequential process of wavefront reconstruction of a beam useful for optical scanning. The subject of holography has been reviewed extensively and elegantly elsewhere (Gab 3, Str 1, Col, Smi, Cau, Lee 2), and its general precepts are coupled with discussions of specialized topics which are cited throughout the book. Introductory representation is offered here to fortify the serious worker with the basics of this prerequisite discipline.

3.5.1. Field Interactions

The electromagnetic field is characterized by two vectors: the electric vector $\mathbf{E}(x, y, z, t)$ and the magnetic vector $\mathbf{H}(x, y, z, t)$. Since photosensitive media (photosensors in general, human vision included) are sensitive only to the magnitude squared of the amplitude of the electric field (the intensity), considerable simplification results from attention to the E field only, which is incident on a photosensitive detector. Further, with optical radiation oscillating at its extremely high frequency (10^{14} to 10^{15} Hz), we can limit discussion to time averages. Thus the intensity is expressed by

$$I(x, y, z) = \langle \mathbf{E}(x, y, z, t)\, \mathbf{E}^*(x, y, z, t) \rangle \tag{1}$$

where the * denotes the complex conjugate (reverse direction propagation) and the angular brackets represent the time average, whereupon the electric field reduces to

$$\mathbf{E}(x, y, z) = \left| \mathbf{E}(x, y, z) \right| e^{i\Phi(x,y,z)} \tag{2}$$

The coefficient is the magnitude term and the $\Phi(x, y, z)$ is the phase term—the fraction of cyclic distance between the source and the point of measurement. Thus $\Phi = 2\pi(\mathbf{r}/\lambda)$, in which the product of the wave number $k = 2\pi/\lambda$ and the distance $\mathbf{r}(x, y, z)$ establishes phase in the direction of propagation, and

$$\Phi = k\mathbf{r} \tag{3}$$

By linear superposition, a point in space receiving radiation from a number p of contributing radiators will experience a total field which is their linear sum:

$$\mathbf{E}(x, y, z) = \sum_p \mathbf{E}_p(x, y, z) \tag{4}$$

The imaging of a distributed object involves the superposition of a near-infinity of points. In the case of optical scanning, the number of points is typically limited to one: a single-point image, often expressed as a "point-source hologram" (Col). When generated on a plane surface, it represents the direct analog to the Fresnel zone lens discussed in Section 3.4.

3.5.2. Dual Source Interference

Let an E field be represented by two coherent waves having amplitudes A_1 and A_2 with phase Φ_1 and Φ_2, respectively, propagating at a temporal phase change

$\omega t = 2\pi f t$, where f is the (common) cyclic frequency and t is the time. The waves, represented by $\mathbf{A}_1 e^{i(\Phi_1 - \omega t)}$ and $\mathbf{A}_2 e^{i(\Phi_2 - \omega t)}$ and incident on a detecting medium will record a time-averaged intensity

$$I = \left| \mathbf{A}_1 e^{i(\Phi_1)} + \mathbf{A}_2 e^{i(\Phi_2)} \right|^2 \tag{5}$$

Expanding, with complex conjugates per Eqs. (1) and (2), we obtain

$$I = A_1^2 + A_2^2 + A_1 A_2 [e^{i(\Delta\Phi)} + e^{-i(\Delta\Phi)}] \tag{6}$$

where $\Delta\Phi = \Phi_1 - \Phi_2$ is the phase difference, appearing only in the product term. Applying identities for the exponential terms, we have

$$I = A_1^2 + A_2^2 + 2A_1 A_2 \cos \Delta\Phi \tag{7}$$

When the amplitudes are equal, the intensity reduces to

$$I = 2A^2(1 + \cos \Delta\Phi) \tag{8}$$

Equation (6) is often represented in conjugate form as

$$I = A_1^2 + A_2^2 + \mathbf{A}_1 \mathbf{A}_2^* + \mathbf{A}_1^* \mathbf{A}_2 \tag{9}$$

The most commonly encountered waves are the most easily generated waves: plane and spherical. The plane wave is expressed by

$$\mathbf{A} = A_o e^{ik\mathbf{r}} \tag{10}$$

while the spherical ones are represented by

$$\mathbf{A}_< = \frac{1}{r} e^{-ik\mathbf{r}} \tag{11}$$

and

$$\mathbf{A}_> = \frac{1}{r} e^{ik\mathbf{r}} \tag{12}$$

where $\mathbf{A}_<$ is diverging and $\mathbf{A}_>$ is converging, $k = 2\pi/\lambda$ is the propagation vector or wave number, and $\mathbf{r}$ is the radial component of the (x, y, z) direction of propagation.

3.5.3. Wavefront Reconstruction

Let the two coherent waves be incident on a material whose recorded transmittance T' is a linear function T of the incident radiation, expressed as $T' = TI$. Reilluminating the material with a reconstruction wave C, an output wave will develop of the form $O = CT' = CTI$. Expanding with Eq. (9) gives

$$O = CT(A_1^2 + A_2^2) + CTA_1A_2^* + CTA_1^*A_2 \tag{13}$$

If the reconstruction wave is made identical to one of the original waves, such as $C = A_2$, in which A_2 is identified as the reference wave and A_1 is now called the object wave, then

$$O = A_2T(A_1^2 + A_2^2) + TA_1 + TA_1^*A_2^2 \tag{14}$$

The first term represents the attenuated reconstruction wave (selected identical to A_2, the reference wave). The second term is the principal one, representing an output wave proportional to the material transmittance and otherwise identical to the original object wave A_1 and continuing its travel in the same direction. The third wave, usually much attenuated, is proportional to the original A_1 but traveling in the conjugate direction. Observation of Eqs. (13) and (14) reveals that to reconstruct a strong A_1^* (oppositely directed) such that the wave re-forms, for example, an original object point in its original location, it is merely necessary to reilluminate the hologram with the (oppositely directed) A_2^*, the conjugate to A_2. Note the discussion at the end of Section 4.2.1.3.1, which accommodates both cases. While the foregoing development exemplifies the use of absorptive media (such as unbleached silver halide) in the region of $dT/dE \approx$ constant, the conclusions are equally germane for phase media in the region where $d\Phi(E)/dE \approx$ constant (Col), providing, in fact, significantly higher diffraction efficiency.

Higher-order waves are normally suppressed by, for example, "thick"-media filtering. A comprehensive discussion of grating thickness and its effect on filtering of the desired reconstructed component is provided in a work cited earlier (Sol 1). When "thin" media are employed, such as photoresist, which forms a phase relief grating with selective dissolution of the surface, efficient reconstruction may be provided by blazing (Sol 1, She 1) or by benefit of formation of a deep grating profile (Loe).

4

Holographic Scan Techniques

4.1. ILLUMINATION-BOUNDARY CONSTRAINTS

4.1.1. Introduction

In this section we express the factors that determine the uniformity of a holographic point image as it rotates about an axis. Only dynamic integrity is considered here: accepting the quality of a fixed image at one point within its scanned range and seeking the relationships between the substrate and its illumination which allow that quality to be sustained during scan.

Since rotational scanners generally maintain a fixed mechanical angular velocity $\dot{\Phi}$, a scanned output beam angular velocity $\dot{\Theta}$ is considered useful when it approaches a linear relationship to the rotation of the substrate, that is, when $\dot{\Theta}/\dot{\Phi} \approx \Theta/\Phi = m \approx$ constant. Optical and/or mechanical means then provide for subsequent linear transformation to image space to make $v_x/\dot{\Phi} =$ constant, where v_x is the spot velocity on the image surface. Straight or circular image locuses dominate for accurate fabrication. Optical methods are exemplified by the flat-field (so-called f-Θ) lens, and mechanical ones by the use of image surfaces that conform to the scan locus. Scan linearity is discussed further in Section 5.2.

Unless indicated otherwise, subsequent discussion applies to rotating holographic scanners operating in transmission, and to those operating in reflection in which the exposing object and reference waves illuminated opposite sides of the substrate. Certain reflection holographic scanners formed by illuminating the same side of a (generally opaque) substrate are given separate consideration in Section 4.1.8.

4.1.2. Radial Symmetry

A basic requirement for certain continuity and uniformity conditions in holographic scanning about an axis is radial symmetry (Bei 1, Cin). We define this

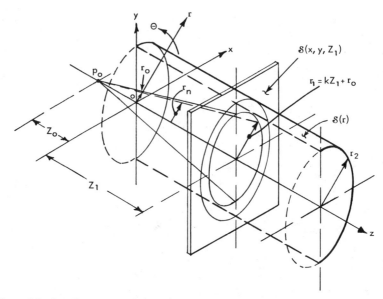

Figure 4.1. Coordinate system for radial symmetry, showing examples of cone and cylinder.

property with the aid of Fig. 4.1, representing rectangular and cylindrical coordinate systems and two intersecting surfaces. One surface $S(x, y, z_1)$ is a plane displaced a distance z_1 from and parallel to the x-y plane. Another is a section of a circular cylinder of radius r_2 extended along z (invariant with Θ). A surface exhibiting radial symmetry is constant in r over varying Θ, allowing only circular cross sections at different z. The right circular cylinder $S(r)$ is a special case (invariant with r for any z). Another is a cone concentric about z (i.e., $r_n = kz_n + r_o$; k = slope, and r_o = radial intercept). Thus, for radial symmetry, the r_n are constant at each z_n. That is, the surface $S(r, \Theta, z) = S(r_n, z_n)$.

An illuminating or reilluminating wave (as well as the rotating body and the scanned output function) can exhibit radial symmetry. Analysis is simplified when assuming that the reilluminating wave incident on the substrate is broad in subtense compared to the extent of a hologram and its displacement. This condition, tantamount to reconstruction with an overilluminating beam (per Section 2.8.1) is useful for understanding and is not constraining in practice. The scanning characteristics thus derived are invariant with over- or underillumination. Also, for these considerations of dynamic integrity, we assume no contributors to stationary aberration (such as wavelength shift or hologram thickness variation). Perturbing factors are reserved for Chapter 5.

4.1.3. Conformal Ray Surface

For further insight, we introduce a diagnostic tool: the conformal ray surface (CRS), a curvilinear sheet of illumination (or portion thereof) interconnecting rays that form radial symmetry, satisfying the surface condition $S(r_n, z_n)$. For a stigmatic spherical wave, the CRS is a regular cone concentric with the axis with its vertex at the geometrical focus. In Fig. 4.1 this is represented by the surface $r_n = kz_n + r_o$, whose geometrical focus is at p_o (displaced by z_o) from the coordinate center o. For a collimated wave, the CRS is a concentric cylinder for which the r_n are fixed at r_2 in Fig. 4.1.

4.1.4. Rotating Disks

Referring to Fig. 4.2, consider a flat disk D mounted normal to an axis and illuminated with a conic conformal ray surface (CRS) converging to (or diverging from) a point P_o, P_o' on the axis. The intersection of this surface with the disk forms a concentric circular arc marked common intersection. Consider the disk rotatable about the axis and sector A rotating clockwise to position A', where its illuminating arc remains identical to that formerly in position A.

Situate within sectors A and A' a pair of identical (point image transmission) holograms such that a common input conjugate point of both appears at P_o'. Grating lines are illustrated linear schematically. The holograms are lenticular, forming off-axis finite conjugate zone lens-type grating structures (Section 3.2). Illuminate with the conic CRS from P_o' and the diffracted image points P_d (of A and A') will be positioned in space at identical radii r_d. Further, they will be

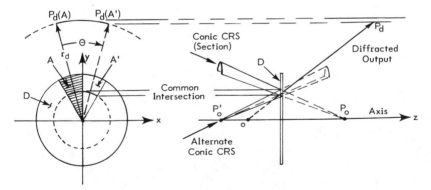

Figure 4.2. Flat disk D normal to axis, illuminated by conic conformal ray surface (CRS). Hologram sector A rotates to A' through angle Θ maintaining stationarity of common intersection and therefore of diffracted principal ray to point P_d with respect to disk.

angularly separated by Θ (in the plane normal to the axis), corresponding to the angular separation of the two holograms in the plane of the disk. With rotation of the (overilluminated) disk about its axis, the holographic segments experience no change in illumination and the image points remain stationary with respect to the disk, rotating with the disk. The output beam then executes a conic scan surface with its apex at o, while P_d traces a circular arc through the angle Θ. Conversely, if the sheet of illumination is not a radially symmetric CRS, its intersection with the holograms is noncircular. Even if the holograms were originally generated with this nonsymmetric illumination, different portions of each hologram will now be illuminated at different positions and different angles during rotation, forming dynamic misorientation of the diffracted scanned image points. Further, the set of reconstructed points from such a (lenticular) hologram will fail to converge at a single point—denoting aberration at all rotation angles except the one at which the hologram was formed.

For the case of a holographic disk reconstructed by a collimated beam (Bei 4, Cin), consider a circular cylindrical CRS concentric about the axis, per Fig. 4.3a. With the cylindrical CRS parallel to the axis (origins at $\pm\infty$), the same situation of stationarity with respect to sectors A and A' exist, a special case of the conic illumination.

If, however, the same cylindrical CRS were directed asymmetric with respect to the rotating axis (Fig. 4.3b), it would not only be intercepted by the flat disk as an ellipse, but rotation of the disk about its axis will deviate the reconstructed point quality and position. This deviation is identical to that resulting from holding the disk fixed and nutating the cylindrical CRS about the disk nodal origin o (maintaining β constant) and monitoring the diffracted reconstruction from a holographic sector. When the hologram is lenticular (exhibits optical power), two principal effects will be observed: (1) aberration due to reillumination over the changing field angle, and (2) misorientation of the reconstructed point due to average grating angular change with respect to the reilluminating beam.

When the hologram is not lenticular but forms a "linear" grating, there will develop no aberration of the diffracted wave during collimated reillumination, for at any instant the angular relationship of all grating regions on the hologram remain uniform with respect to the illuminating beam. However, misorientation will develop, for while this uniform relationship exists at any instant, it varies dynamically during scan. This could affect the output angle adversely or usefully, as discussed in subsequent consideration of asymmetric illumination. Also, during asymmetric illumination, we need contend with the additional effect of changing diffraction efficiency due to this dynamic change of the relative angular relationship of the reillumination wave and the hologram.

Grating lines are illustrated schematically in Figs. 4.2 and 4.3 as tangential to the disk. For linear gratings, this denotes the most prevalent orientation, in

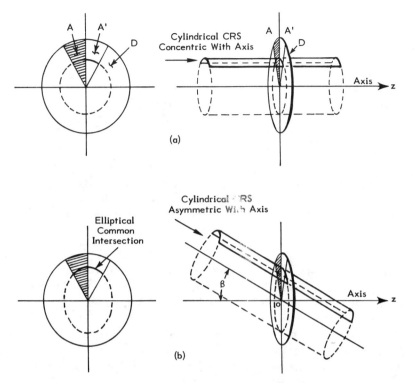

Figure 4.3. Flat disk illuminated by cylindrical CRS. (*a*) CRS concentric with axis for circular common intersection. (*b*) CRS asymmetric with axis for elliptical common intersection.

which the principal ray of the diffracted beam remains radial (e.g., per r_d in Fig. 4.2) and the scan locus is generally tangential. Application exists in which the central grating lines are in quadrature (i.e., radially oriented) and the output scan locus is also generally radially oriented. This interesting special case is discussed in Section 5.1.2.1.

4.1.5. Rotating Cylinders

A right circular cylinder can be considered as assembled from a continuum of circular rings that are concentric with the axis. Cylindrical sections with two bounding rings are illustrated in Fig. 4.4, in which two illuminating options are shown. One is a spherical wave, as derived from or to a point source (Pol 1) for transmission or for reflection (Ish 2, Ish 3) and the other is a cylindrical wave developed from a (coherent) line source (Bei 7).

Taking the spherical wave condition first, consider Fig. 4.4*a*, which illus-

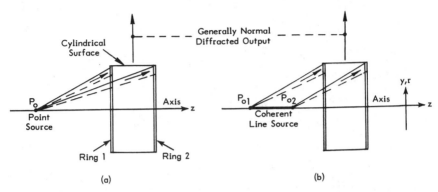

Figure 4.4. Cylindrical substrate bounded by rings 1 and 2. (*a*) Illuminating CRS derived from point source. (*b*) Illumination from CRSs of coherent line source. Principal ray of diffracted output can be normal to the axis.

trates bounding concentric rings 1 and 2 illuminated from two different CRSs, both derived from the same axial point p_o. These thin rings can be considered to intercept a spherical illuminating wave, each acting as was the disk in Fig. 4.2. Since the entire system is radially symmetric, the relationship between a full set of CRSs and the full cylinder is invariant with rotation. Hence, so illuminated, the characteristics of a diffracted output wave will be invariant with rotation. The principal ray of the output wave can be designed to radiate in a direction generally normal to the rotating axis.

The second case of illumination from a coherent line source is represented in Fig. 4.4*b*, again for the two bounding concentric rings. The two CRSs are here derived from two different (mutually coherent) sources, p_{o1} and p_{o2}. The illumination needed for this arrangement is that derived from, for example, an axicon (Edw 1, Edw2, McL) forming a line focus coincident with the axis. As for the spherical case, the two thin rings can be considered to intercept the respective CRSs. Thus the active cylinder, when illuminated with a full set of such CRSs, will remain symmetric and cylindrical. The hologram can also be made to diffract generally normal to the rotating axis.

Both forms of illumination provide rigorous reconstruction. That from a single point (spherical wave) is usually simpler to generate, while that from the line source has two virtues: (1) the output is independent of small axial misorientation, and (2) construction and reconstruction conjugate waves are identical in certain reflective systems, subsequently discussed. The reflective cylinder and cone (Sections 4.1.8.3 and 4.1.8.4) exposed with object and reference waves on the same side of the hologram require a line image as formed by axicon illumination.

With the allowance of these two options, it may be concluded that others exist. In fact, a range of illumination sources and directions is allowed. While the simplest to generate is the point source and the next likely is the coherent collimated line source exemplified above, there is no restriction on the degree of collimation. The wave may be diverging or converging in the r-z plane. That is, with the ray slope designated as $k = dr/dz$, if ray a in one CRS is expressed by $r_a = k_1 z_a$, then a set of n rays in the r-z plane is collimated when $r_n = k_1 z_n$, and is converging or diverging when $r_n = k_m z_n$ and the k_m form a monotonic and unidirectional change of slope. In the r-z coordinate system of Fig. 4.4b, when dk/dz increases with z_1, it is converging. The restriction on the integrity of the illuminating function is that it be derived from a line coincident with the axis (a set of valid CRSs) and that it be reproducible in a practical manner from an axiconal optical system (Bei 7).

4.1.6. Rotating Cones

The axially symmetric cone (Fig. 4.5) is an interesting variation (Bei 3, Kra 3). It exhibits the characteristics expressed for the rotating (transmissive) cylinder, with the added option of allowing illumination from a paraxial collimated beam (from a point on the axis at $-\infty$) while the diffracted output (principal ray) may be directed essentially normal to the axis. This special case is illustrated in Fig. 4.5. As in all previous cases, this also allows for operation in reflection by reconstructing with the conjugate to the reference wave, considered thus far only for holograms on transparent substrates exposed from opposite sides.

4.1.7. Rotating Spheres

The concentric spherical section is unique, for it exhibits not only axial symmetry but also point symmetry (Bei 2). As is well recognized in the optical

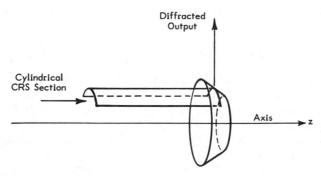

Figure 4.5. Conic substrate illuminated with (paraxial) collimated CRS. Principal ray of diffracted output can be normal to axis. Options for illumination from point or line sources as expressed for cylindrical substrate also apply.

fabrication community, the spherical surface is simplest to generate accurately, for it requires control over only one degree of freedom: its radius. In transmission (or in reflection when the hologram is exposed from opposite sides), it exhibits all the characteristics of illumination expressed above, while maintaining radial symmetry.

A narrow disk sliced from a sphere can be approximated by a section of a cylinder or cone, depending on the portion of the sphere utilized, and its illumination requirements can be adapted directly from earlier discussion. The next section addresses the holographic scanner operating in reflection, when exposed originally from the same (outside) surface, typically on an opaque substrate.

4.1.8. Reflection Holographic Scanners Exposed from the Same Side of the Substrate

4.1.8.1. Introduction

The discussion thus far applies to scanners in which the exposing reference wave was derived *from* the axis. For transmission holograms, the object wave is also incident on the same side. For reflection holograms on a transparent substrate, the object wave is incident on the opposite side of the substrate. Reillumination for transmission effectively duplicates the reference wave, while reillumination for reflection is directed from the *opposite* side, forming the conjugate to the reference wave. That is, if continued beyond the substrate, it would propagate in a direction precisely opposite to that by the reference wave, and converge *toward* the axis.

When, however, a reflection hologram is required needing formation from the same (outside) side of the substrate (such as when the original substrate is opaque), special consideration is necessary to generate a readily realizable reconstruction wave. This requirement is most prominent for a curved surface, subsequently discussed, followed by the same (but simpler) procedure on a flat substrate.

4.1.8.2. Reflective Spheres and Ellipsoidal and Parabolic Variations

When a spherical substrate is holographically exposed with both waves from the "outside," point symmetry is shown to provide for practical reconstruction (Bei 1). This is illustrated in Fig. 4.6, which shows the rotational spherical surface (meridional section) intended for operation in reflection. To reconstruct a reflection hologram (exposed from the same side) stigmatically, it is necessary to reilluminate with the conjugate of the original reference wave—that is, that one which, when redirected into the system, reforms the reference wave (or portion thereof) in precisely the opposite-going direction.

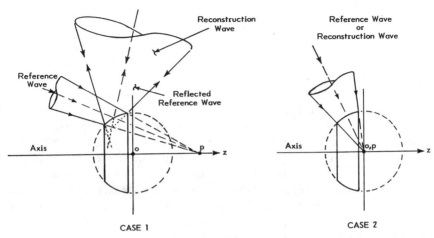

Figure 4.6. Spherical substrate sections in holographic construction and reconstruction in reflection. Case 1 requires reconstruction with severely aberrated wave to form valid conjugate, while in case 2 the reference and reconstruction waves are identical and self-conjugate. Object and diffracted output waves, typically normal to the axis, are not shown, for simplicity.

The reference wave used to form the hologram is shown in two cases. Although both are valid portions of much steeper cones which are radially symmetric with the axis (as CRSs), the reference wave in case 1 intersects the axis at point p, which is displaced from the center of the sphere, while in case 2 the axis is intersected at the center of the spherical section, o. The conjugate wave needed for stigmatic reconstruction of the reflection hologram is shown as an extremely complex function in case 1 and a uniquely simple one in case 2. The reconstruction wave is developed for case 1 to form the conjugate to the reference wave (after reflection from the spherical surface), where it is shown to be severely spherically aberrated, manifest by the (dotted) extensions to the (three) rays within the substrate which form no common intersection. It also exhibits a steep converging angle, requiring formation with an optical system having an exaggerated high numerical aperture.

For case 2, the reconstruction beam is identical to the reference beam. When the center of the sphere is also the point of illumination symmetry, all rays are normal to the local surface and the same stigmatic reference and reconstruction beams form self-conjugates.

To generate complex illumination functions, as manifest for case 1 and for several other conditions subsequently described, a general solution is to form a holographic optical element (HOE) "objective lens" by employing the complicated reflected reference wave as a new object wave and a simple wave as a new reference wave on the substrate to become the new objective lens. The complex function may be reconstructed, thereby, from simple illumination. However, optical efficieency,uniformity, and alignment accuracies need to be considered. Holographic optics to serve as objective lenses for scanning applications were described in 1971 (Bei 1, Dar).

Innovative variations to the spherical surface allow exposure and reconstruction with longer-focal-length spherical waves (having lower numerical apertures) (Kle 2). Referring to Fig. 4.6, case 1, if the substrate were an ellipsoid with its major axis on the rotating axis and point p one of its foci, formation of the hologram with a spherical reference wave directed to the conjugate focus would provide stigmatic reconstruction. The use of a central sector of this ellipsoidal surface is particularly attractive to approach the characteristics of the next-described cylindrical configuration, while maintaining spherical reference and reconstruction wavefronts. Another variation is a paraboloid whose axis is on the rotating axis, allowing formation of the hologram with a reference beam directed to its focus and reconstructed with a *collimated* wave parallel to the axis. The roles of the reference and reconstructed beams can be reversed. The principal limitation to these clever variations is that it becomes necessary to form precise aspheric substrates (ellipsoidal and parabolic), while the original system was formed on a spherical substrate, deriving intrinsically high fabrication accuracy from simple polishing facilities and processes.

4.1.8.3. Reflective Cylinders

The cylindrical scanner exposed from the "outside" also utilizes a unique illumination function to allow the use of identical reference and reconstruction waves (Bei 7). This is illustrated in Fig. 4.7, which shows a cylindrical surface (meridional plane) illuminated in two conditions: (1) from the left (fine lines) with a reference wave during holographic exposure with a generally normal object wave and (2) from the right (bold lines) with a reillumination wave for reconstruction. Both illuminating functions are identical and self-conjugates, derivable from the same axiconal objective lens mounted alternately (radially) symmetrically on either side. As developed for the cylinder operating in transmission per Fig. 4.4b, all illuminating rays (in the r-z plane) make an equal angle with the axis and are directed to form a coherent line image on the axis. Thus the same wave-shaping procedure of the reference beam can be used to form the reconstruction beam. Conversely, any other well-ordered beam used for the reference wave (e.g., spherical) requires generation of a rather complex reconstruction wave. In principle (as is that for the spherical spinner), such an aberrated wave may be formed by creating a holographic optical element to serve as its shaping device, whereby the reference wave as reflected from the substrate serves as the object wave of the holographic objective.

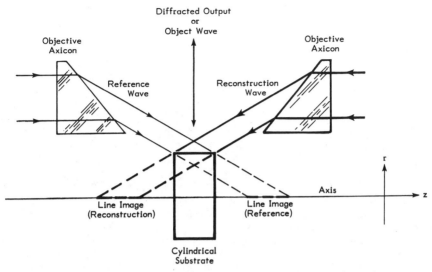

Figure 4.7. Cylindrical substrate (meridional plane) in holographic construction and reconstruction in reflection. Same objective axicon (radially symmetric about the axis) can serve both conjugate functions.

4.1.8.4. Reflective Cones

The conic substrate exhibits the illumination characteristics expressed for the cylinder, except that the object and reference waves will exhibit differing angles with respect to the axis. This requires that the axicons which may serve as reference or reconstruction objectives need to be different components. All the other properties expressed for the reflective cylinder apply to the reflective cone. An interesting condition develops when the cone is at 45° to the axis and is illuminated during exposure or reconstruction with a paraxial collimated wave. The conjugate to this is a cylindrical wave (as derived from a cylindrical lens) arriving at all points normally on the axis. This wave is to be oriented, however, so that it avoids occupying the same object/image space of the object or reconstructed waves. Another special case exists, illustrated in Fig. 4.8, which not only avoids these interferences, but allows reconstruction with a simple collimated wave. The reference wave angle β (from a radially symmetric axicon) and cone angle α are so chosen ($\beta = 2\alpha$) that the reflected reference wave is paraxial and collimated. This permits use of a correspondingly simple wave for stigmatic reconstruction (Bei 3).

4.1.8.5. Reflective Disks

While the cylindrical, conic and spherical substrates supporting reflective holograms needed special attention for proper reillumination, the reflective disk

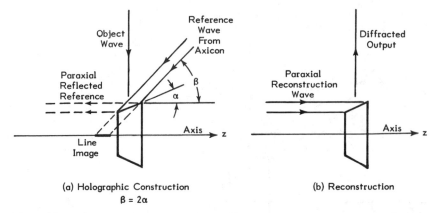

Figure 4.8. Conic substrate (meridional plane) during (a) Holographic construction and (b) reconstruction in reflection. When $\beta = 2\alpha$, paraxial reflected reference wave requires simple paraxial collimated reconstruction wave for valid conjugate. Only principal rays of object and diffracted output are shown.

scanner exhibits some of the same options of illumination as in transmission (Bei 4, Bei 5). This is due to its (nominally) planar surface, which acts as a perfect (zero-order) plane mirror, allowing the generation of conjugates which are either spherical or collimated waves. Figure 4.9 illustrates the spherical wave configuration, in which the reference is converging and the reconstruction is diverging. These roles may be reversed. The collimated wave case (discussed and illustrated in Section 4.2.2.2) is self-conjugate, as was the spherical wave for the spherical substrate (Fig. 4.6; all rays normal to the substrate or wave-

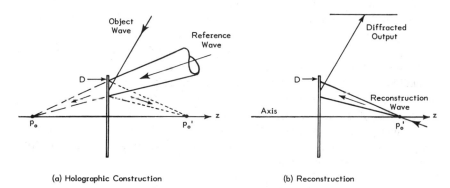

Figure 4.9. Disk substrate D (meridional plane) during (a) Holographic construction and (b) reconstruction in reflection. Point p_o' is mirror image of p_o and source of reconstruction wave for valid conjugate. Only principal rays of object and diffracted output are shown.

fronts coincident with the substrate surface). Another form of disk-like scanner utilizing this spherical wavefront conjugate condition is the concave configuration discussed in Section 4.2.2.2.1. A notable case of noncompliance with self-conjugate reillumination follows in Section 4.2.2.2.2, with regard to the work of Locke (Loc) on a reflective disk scanner formed on a silver halide plate (Rig). This may be likened to the operation of a transmission hologram having one mirrored surface.

4.1.9. The Plane Linear Hologram

Most of the considerations provided earlier relate to the *lenticular* hologram— one exhibiting optical power in transforming an object point to a conjugate image point. Because of the intolerance of such holographic optical elements to field angle misorientation, reillumination demands precise conformance (e.g., congruence) of the reconstruction wave with the original reference wave; and rotational dynamics requires radial symmetry. One can, of course, abrogate radial symmetry to suffer (and perhaps correct for) the spot abberation and locus deviation for the sake of some other potential benefit (Lee 1), as expressed in Sections 2.8.4 and 4.2.2.6.

The linear hologram, sometimes identified as a plane linear diffraction grating (PLDG), is a periodic array of mutually parallel straight grating lines on a plane surface. Thus the main rotational configuration that will support a perfectly linear hologram is the flat disk. One might offer two departures, both on a cylindrical substrate: On one, the rulings are parallel to the axis, and on the other, they are normal to the axis. In both cases, however, they provide no scan function, for in effective translation, they exhibit no diffractive change with respect to a fixed illuminating beam during rotation of the substrate. The linear hologram can scan only by angular change of the grating lines with respect to the illumination. Such configurations are discussed extensively in Section 4.2.2.

4.2. ROTATIONAL SCANNER DEVELOPMENT

Following the expression of some of the fundamentals in Section 4.1, we now address the characteristics of systems accorded substantive analysis and/or experiment. The discussion is divided into two major parts represented in Sections 4.2.1 and 4.2.2, depending on whether or not the rotating substrate provides a surface projection normal to the axis. Those that provide this normalcy are intrinsically able to provide straight-line scans; while those that do not, having generally disk-like substrates, undergo special operations or adjustments to achieve straight-line scans.

Historic progression (Appendix 1) indicates that the very earliest holographic scanners were considered for disk-like substrates (Bei 4, Cin, Gab 1, McM 1, McM 4). It was then appreciated, however, that this form was destined for limited service unless something was done to develop straight-line scans—a major subject in its own right, discussed in Section 4.2.2. Thus substantive attention was directed to the techniques classified in Section 4.2.1: those having surfaces that project normal to the axis.

4.2.1. Scanners Providing Surface Component Normal to the Rotating Axis

4.2.1.1. *Reflective Holofacet Scanners*

The first patented holographic scanner was entitled "Light Scanning System Utilizing Diffraction Optics," granted on October 19, 1971 (Bei 1). Following extensive testing, it was reported in 1973 as the Holofacet scanner (Bei 2, Bei 4), its purpose being to provide combined resolution and speed which remains to this date unsurpassed by any single-channel optical scanner, conventional or holographic. Motivated by development for high-performance reconnaissance image scanning and recording, it executed scans at 10,400 per second, covering 20,000 elements per scan at 200 megapixels per second. This Holofacet Optical Scanner Apparatus received the 1973 (Industrial Research) IR-100 Award and is now in the permanent collection of the Smithsonian Institution in Washington, D.C. The substrate is a polished 2-in. radius beryllium sphere mounted on a shaft, coupled to a two-phase induction motor and driven at 52,000 rpm. The optical configuration followed the principles expressed in Section 4.1.8.2, as

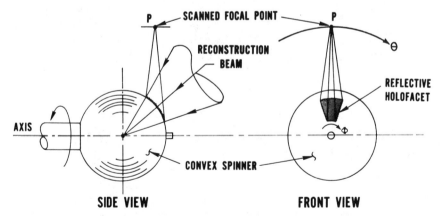

Figure 4.10. Reflective Holofacet scanner, overilluminated. Reconstruction beam converges to center of spherical spinner. Diffracted output beam forms focal point *P* which is scanned on arc concentric with axis. Alternatively, a collimated output beam may be focused by subsequent objective lens to flat field.

illustrated in Fig. 4.6, case 2. Figure 4.10 represents further the optical config-uration, while photographs of the spinner-lens assembly, the optical exposure assembly, and the recording test facility appear in Figs. 4.11, 4.12, and 4.13, respectively.

Figure 4.10 illustrates this radially symmetric reflective system. The recon-struction beam (side view) is directed to a focal point that is coincident with the center of the spherical substrate. The diffracted output beam converges to point P and executes scan Θ coincident with substrate rotation Φ. Holographic exposure is accomplished with a reference beam identical to that for reconstruc-tion—all rays are normal to the surface, hence self-conjugates in reflection, while the object beam is directed from point P to the substrate. Exposure and reconstruction were with an argon-ion laser at 488-nm wavelength.

In the configuration tested, 2 of the 12 Holofacets were overilluminated (p. 26, Section 2.8.1) by the reconstruction beam. Thus the entire aperture con-tributes to the $f/6$ output cone during the full $360°/12 = 30°$ scan angle. Under these conditions, the duty cycle is unity (retrace interval is zero). As with conventional polygons, underillumination is also allowed, providing the same options with this and other holographic scanners, whereby illumination efficiency is raised to (near) unity, the duty cycle is depleted by $\Delta = D/D_{max}$ (per Section 2.8.1), and the substrate must be increased in diameter by approx-imately $1/\Delta$ to maintain active aperture size. This underilluminated form is illustrated in Fig. 4.14, which shows a narrow Holofacet element designed for

Figure 4.11. Photograph of Holofacet spinner–lens assembly. From left to right, drive motor, flexible coupling, removable Holofacet sphere, and reference/reconstruction objective lens.

Figure 4.12. Photograph of Holofacet exposure facility. Microscope objective lens in upper center forms object point and object wave, while reference beam from (diagonal) expander illuminates objective lens.

Figure 4.13. Photograph of Holofacet recording test facility. Clockwise on table: laser, folding mirrors, wideband E-O modulator, beam expander, and spinner–lens assembly. Broadband modulator driver and test apparatus at right.

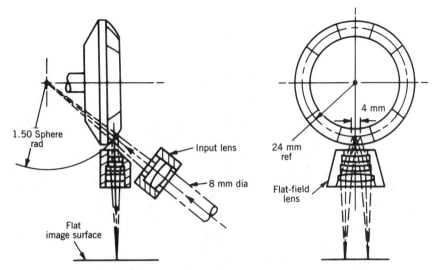

Figure 4.14. Reflective Holofacet scanner, underilluminated. Designed for microimage scanner (at 100 lp/mm) on flat surface. (From Ref. Bei 7.)

flat microimage scanning or recording. The collimated output beam is focused by a flat-field (f-Θ) lens to a 5-μm spot over a microimage format. The same number of elements can be transferred to a larger standard page format with appropriate increased focal length of the f-Θ lens. As with the overilluminated form, it exhibits the same functional characteristics of a pyramidal polygon (Bei 6) while maintaining ease of accurate fabrication and control of the one spherical surface.

The processing of the large solid substrate for reflection holography merits expression for its general significance. A precision (uniformity to 2.5 μm before polishing) beryllium sphere was machined to provide flat sides and a through-hole designed for shrink-fit around a steel shaft as in Fig. 4.15, forming journals for subsequent mounting for processing and for high-speed ball bearings. Twelve equispaced pilot holes were oriented on the flat sides of the substrate for facet indexing during holographic exposure. Since this is an opaque substrate upon which exposing reference and object waves are to arrive from the same side, the beryllium surface was made optically absorptive by means of black anodizing (approximately 8 μm thick) to prevent destructive interference of either incident wave by a partially reflected wave within the thin photosensitive coating. The anodized surface was then polished to provide a $\frac{1}{10}$-wave-per-inch uniformity over the band representing the facet locations. Each of the 12 facets subtends approximately 1 inch. As is appreciated from the fabrication practices of conventional optics, a spherical surface is the most efficient for precision

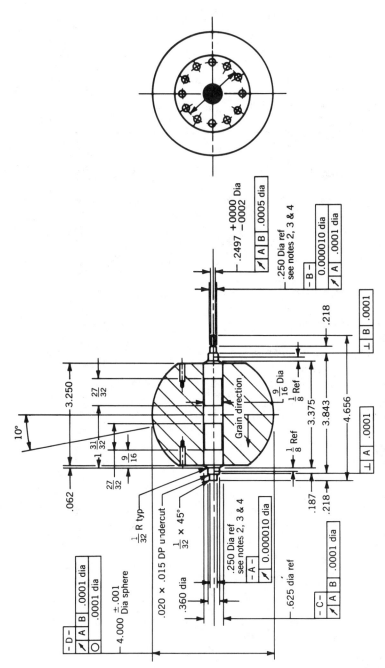

Figure 4.15. Manufacturing drawing for beryllium substrate and shaft. Journals provided for bearing mounting and demounting during exposure, processing, and test. Twelve equispaced pilot holes provide facet indexing during holographic exposure.

70

optical polishing, motivating adaptation of that curvature. Beryllium was selected to take advantage of its high specific stiffness ($m_{ss} = E/\rho$, where E is Young's modulus of elasticity and ρ is the material density) for low deformation during high-speed rotation (Bei 6).

The photosensitive coating was photoresist—positive-working Shipley AZ-1350H, applied by spraying a dilute solution on the sphere while rotating on a turntable. The dilution ratio of acetone to resist was extremely high, 90:1, so that upon evaporation a very thin photosensitive film remained.

Exposure was performed in the manner represented in Fig. 4.12; the 12 facets were exposed sequentially upon indexing to the corresponding pin-locked pilot holes. After exposure, the spinner was removed from its bearing assembly, the resist processed, and the exposed band aluminized to provide high reflectivity. Although the theoretical diffraction efficiency of reflective sinusoidal phase gratings is 33.6% (Urb 1), due to partial blazing (She 1) of the thin film of photoresist, efficiency was measured at approximately 42%. No attempt was made at that time to approach the theoretical maximum of 100%.

4.2.1.2. Variations in Reflective Holofacet Scanners

Several variations of this configuration were described (Bei 1, Bei 7, Bei 10). One provides for the diffracted beam to be extracted from the maximum diameter portion of the spherical surface, achieving symmetric output from a minimum-diameter spinner (for the same aperture size and number of facets). This allows near-cylindrical operation of the spinner, illustrated in Fig. 4.16. Normal incidence of the reference/reconstruction rays upon the substrate is provided by illuminating collimated light on a parabolic reflector segment having a slit for passage of the output beam to the scanned focal point. A similar ellipsoidal reflector segment would allow the option of illumination with a diverging wave emanating from its conjugate focal point, effectively conserving the second stage of a beam expander. Alternative illumination equivalent to those from the parabolic or ellipsoidal sectors can be provided by HOEs (Bei 1).

This symmetric configuration is adaptable to provide uniform multiplexing of several data channels from one scanner member. Figure 4.17 illustrates a four-channel arrangement in which crosstalk is suppressed by allocating individual diffractive bands or rings to each channel and ensuring that each input beam illuminates its band and none other. This allows independent encoding of each channel. Multiplexing is discussed in Section 4.4.

4.2.1.2.1. Solid Cylindrical Holofacet Scanners

A major variation is realized in converting to a true cylindrical spinner shape (Bei 7) represented in Figs. 4.18 and 4.19. The solid cylinder not only provides more uniform inertial stress throughout the rotating member (due to elimination

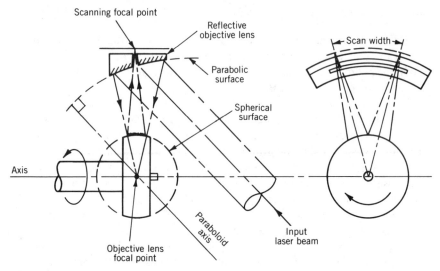

Figure 4.16. Reflective Holofacet scanner providing near-cylindical substrate operation. Parabolic reflector forms objective lens having slit for transport of scanned diffracted beam to image surface. A holographic optical element may serve as an objective lens.

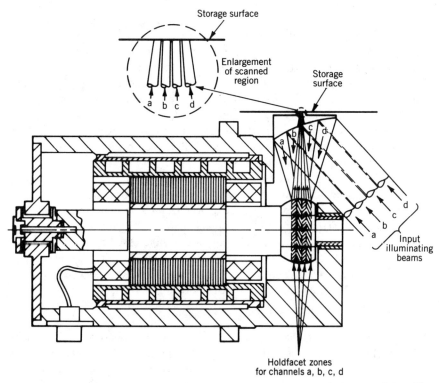

Figure 4.17. Multiplexed Holofacet scanner, formed from prototype of Figure 4.16. Provides independent scan or superimposed on one scan line for, for example, color recording.

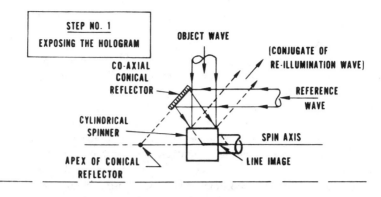

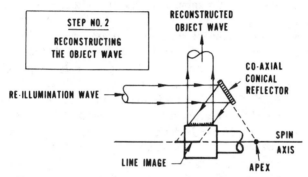

Figure 4.18. Cylindrical Holofacet scanner with axicon objective lenses formed by (same or similar) coaxial conical reflectors. Step 1: exposing the hologram; step 2: reconstructing the object wave.

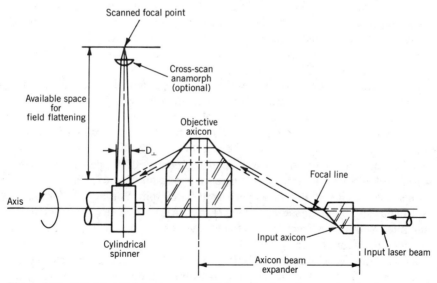

Figure 4.19. Cylindrical Holofacet scanner utilizing refractive axicons and optional cross-scan anamorphic correction (see the text).

73

of change of radius along its surface), but allows extraction of the output beam without physical interference with the reconstruction beam optics. As described in Section 4.8.3, for radial symmetry the reflective cylinder exposed from the outside requires a reference (or reconstruction) line image, generally derived from axiconal optics. Figure 4.18 illustrates reflective and Fig. 4.19, refractive axiconal elements. The same axiconal optics may be applied in both holographic exposure and reconstruction, as was expressed for Fig. 4.16, taking advantage of their mirror-image self-conjugate relationship. Figure 4.19 shows a more complete optical set, consisting of a refractive axiconal "beam expander" (allowing a small input collimated beam), whose second element serves as a combined "objective lens" for spinner illumination, again following the discussion relating to Fig. 4.16.

While comprehensive discussion of anamorphic beam handling for cross-scan error reduction is reserved for Section 5.1, one form of this technique, described in 1974 (Bei 7) is noteworthy here. As represented in Fig. 4.19 a single cylindrical lens is indicated in the image region as a field lens. The diffractive cross-scan aperture $D_\perp$, which is made smaller than normally required, launches illumination which is intercepted by the anamorph to converge the beam as though it was derived from a larger aperture D, corresponding to that required for the final spot size. The cross-scan positional error is reduced by the factor $D_\perp / D$, as described in Section 5.1.2.

To ensure reconstruction of a nominally round (isotropic) and stigmatic final spot, during holographic exposure the object wave is directed to propagate conjugately through the focal point (fixed on the scan surface), whereupon the wave continues back through the cylindrical lens to the photosensitized spinner surface, where it is intercepted by the reference wave to form the facet hologram. Upon reconstruction, the diffracted wave will then propagate as illustrated in Fig. 4.19 and converge properly on the image surface. Observe conjugate orientations as in Fig. 4.18.

An unusual configuration is represented more operationally in Fig. 4.20. It was designed for ultrahigh performance, motivating the use of the stable cylindrical hologram substrate composed of a solid material such as beryllium. Surface processing techniques are discussed in Section 4.2.1.1. Design characteristics as follows: scan format = 9-in. line length (228.6 mm); line rate = 10,000/sec; spatial resolution = 110 linepairs/mm at 50% response (spot size ≈ 4 μm FWHM derived from $f/6$ cone at $\lambda = 442$ nm (Bei 6)). For six Holofacets, rotor radius = 30 mm at speed = 100,000 rpm. For eight Holofacets, rotor radius = 50 mm at speed = 75,000 rpm. Air bearing suspension system configured with optical section isolated at low gas pressure to reduce aerodynamic drag and turbulence.

To place this capability into functional perspective, when compared to the system that was tested to the highest performance on record (see Section

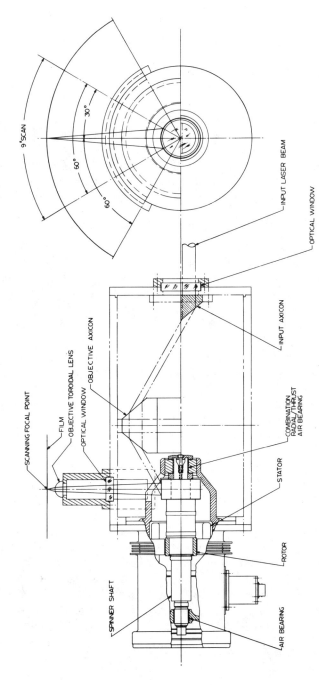

Figure 4.20. Operational design for ultraimage cylindrical Holofacet scanner providing 50,000 elements/scan at 500 megapixels/sec.

75

4.2.1.1): 20,000 elements/scan at 200 megapixels/sec, this image scanner-recorder was designed to provide a $6\frac{1}{4}$-fold advantage in product of resolution and speed: to 50,000 elements/scan at 500 megapixels/sec. Contrasting a typical reprographics scanner which may provide 3000 elements/scan at 10 megapixels/sec (Bei 10), it provides the resolution–speed product of 830 such scanners—thus the rather elaborate configuration represented in Fig. 4.20, in which the need for reduced air pressure on this aerodynamically favorable design is untested. The Holofacet system described above (4 in.-diameter spherical spinner at 52,000 rpm) was tested in air at normal pressure.

Other variations for the reflective (solid member) diffractor having a surface component normal to the axis are the ellipsoidal and parabolic substrates and the conic substrate, discussed in Sections 4.1.8.2 and 4.1.8.4, respectively.

4.2.1.2.2. Cylindrical Form CGH Scanners

Comprehensive work in flattening the field of a cylindrical form reflective scanner was conducted by Ishii (Ish 2, Ish 3), in which the operating holographic facets were exposed through an optimally determined intermediate computer-generated hologram (CGH). To ease the rather complex CGH formation and reexposure processes, the spatially variable image detail was calculated on the basis of flat hologram facets. Experiments were then conducted with a four-faceted regular (square) polygon, assembled using computer-synthesized flat holograms (interferometrically generated, operating in reflection). Radial symmetry was retained, using reillumination in a manner similar to that depicted in Fig. 4.14. While this work yielded some significant results, it is expected that a real operational system need be formed in a circular cylindrical configuration rather than from an array of flat holographic facets which retain all their inertial and aerodynamic limitations. Thus we are left with the demanding requirement of transforming the computation, CGH generation, holographic exposure, and assembly in a circular format. This may be aided by leaving everything in the flat domain until the final ''wrapping'' of the holograms on a drum mandrel (maintaining tolerable surface deformation)—nontrivial, for the computation must be adapted from curved to flat so that when recurved, it is correct. Also, as in any such step-and-repeat procedure, it must be instrumented with adequate precision such that the holograms (perhaps replicated on a strip) exhibit correct relative orientation and ''close'' properly; that is, when the end of the last hologram joins with the start of the first, it maintains accurate matching orientation.

These practical considerations need not detract, however, from the noble effort expended and good results achieved. A 6-mm-diameter beam was focused by the hologram and scanned over a flat image surface while retaining approximately $\frac{1}{4}$-wave spot integrity over $\pm 14°$ ($\approx \frac{1}{2}$ rad). The diffraction-limited resolution per Eqs. (28a) and (28b) of Section 2.8.4 using given values of Θ $\approx \frac{1}{2}$ rad, $D = 6$ mm, $\lambda = 515$ nm, $m = 1$, $r = 80$ mm, and $f = 470$ mm and estimating $a = 1.25$ at 50% MTF for a uniformly illuminated round aperture (Bei 6) reveals $N = 5450$ elements/scan. Since their performance provided approximately $\frac{1}{4}$-wave aberration spot quality, this corresponds to $\approx 20\%$ reduction in resolution, resulting in $N_{eff} = 4540$ elements/scan—very respectable, indeed, for such a system. Furthermore, the optimizing form of the program was utilized to balance good spot quality with reasonably good scan linearity on a flat surface.

4.2.1.3. Transmissive Scanners with Surface Component Normal to the Rotating Axis

4.2.1.3.1. Transmissive Holofacet Scanners

The Holofacet scanners described in Section 4.2.1.1 were configured to operate in reflection from solid beryllium substrates, to maximize inertial stability during extremely high speed rotation. There is, however, no limitation to operation in transmission under more relaxed conditions, maintaining the advantage of accurate fabrication of spherical surfaces and the input rays retained normal to these surfaces. Thick transmission (glass) substrates were described in 1969 (Bei 18). One is represented schematically in Fig. 4.21 to provide a large scanning aperture and to accommodate demanding inertial conditions. The original design, introduced in Appendix 1 and illustrated in Fig. A1.8, shows an outboard bearing assembly for added stability. Figure 4.21 here shows a cantilever

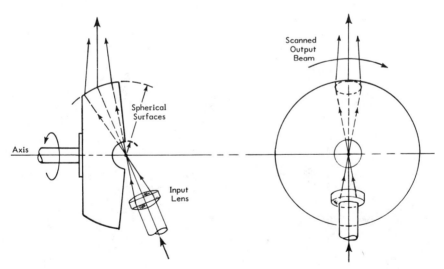

Figure 4.21. Transmissive Holofacet scanner utilizing solid (glass) substrate to constrain inertial deformation during large-aperture output at high speed.

arrangement for illustration simplicity. Where this may be relaxed even further and performance objectives allow reduced inertial loads, a thin-walled glass substrate may be used, as recently described (Bei 10). Figure 4.22 illustrates this configuration in a flat-field operation typical of that for business graphics and graphics arts. By taking advantage of some properties of the transmission hologram, mechanical integrity may be relaxed over that required for the reflective configuration. This takes the form of reduced sensitivity to substrate wobble under conditions of (or approaching) equal illumination and diffraction angles (with respect to the substrate normal). Discussion of these and related factors of compensation for wavelength shift (upon exposure at one wavelength and reconstruction at another) appear in Chapter 5.

Although the two transmission systems described above appear similar functionally during beam reconstruction, they require differing hologram formation so that they may be exposed from the same side of the substrate. Because of

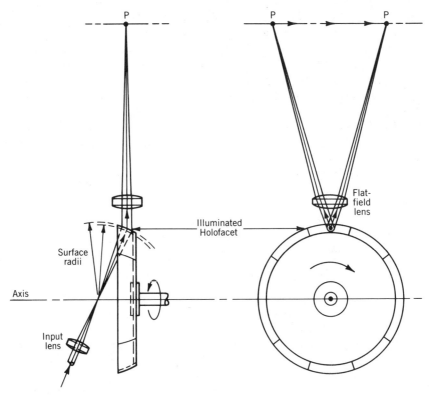

Figure 4.22. Transmissive Holofacet scanner for business graphics and graphic arts resolution and speed applications. Flat-field lens focuses collimated output beam to point P for straight-line scan.

interposition of the thick glass substrate, the first method represented by Fig. 4.21 requires illumination of the object and reference waves from the "outside," both as conjugates of the image and reconstruction waves. The thin-walled method represented by Fig. 4.22 may be exposed more conventionally from the "inside," with the use of a mirror to redirect the collimated object wave normal to the axis toward the substrate (Kuo). In both cases the reference and reconstruction waves are spherical. See the discussion in Section 3.5.3 providing a theoretical basis for direct and conjugate illumination techniques.

4.2.1.3.2. Transmissive Cylindrical Scanners

As the spherical scanners were transformed to transmissive, so may be the cylindrical (and near-cylindrical) ones described in Section 4.2.1.2. Conditions for cylindrical illumination in transmission are expressed in Section 4.1.5.

A notable transmissive cylindrical holographic scanner is that described by Pole and Wallenmann (Pol 1) in 1975 and configured for test as a document scanner (Pol 3) in 1978. Reilluminated from a point on the axis in Fig. 4.4a, it exhibits radial symmetry. Figure 4.23 illustrates this implementation. Considering the solid-line components of the illustration first, the substrate is a thin, hollow glass cylinder on whose inside surface are prepared a series of almost square (25 mm × 26 mm) equally spaced "Hololenses" which image scanned

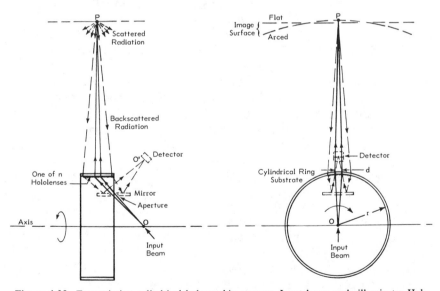

Figure 4.23. Transmissive cylindrical holographic scanner. Input beam underilluminates Hololens, which focuses diffracted beam to image surface. Dashed lines designate optional system for collection of backscattered radiation for document scanning.

point P conjugately from input point O on the axis. The output beam diffracted by the hologram is scanned in a plane generally normal to the axis. When underilluminated (as illustrated), the diffracted beam exhibits the relatively high f-number required for the desired ($\approx 45 \ \mu m$) spot size at P. When the cylinder is rotated, the diffracted beam executes the scan function illustrated in Fig. 2.7 and detailed in Section 2.8.3 to be comprised of the regular scan angle Θ plus the augmented angle α (when hologram displaced by r focuses over a distance f). Ideally, the circular scan locus should be matched by the image surface along the arced path to maintain perfect focus. With the high f-number, however, the flat compromise shown was found to be acceptable within the allowed focal depth over the relatively small total scan angle of approximately $25°$.

The dashed lines in Fig. 4.23 denote an optional utilization of the reciprocal properties of the system. The full hololens aperture subtends a backscattered component of the scattered radiation from the image surface. The returning component is rediffracted and effectively descanned by the scanner and redirected toward O. It is intercepted by a mirror to its image O', where a detector is located to derive therefrom a signal representing the backscattered flux, used for image digitizing. Retroreflected signal detection (retrocollection) is not unique to holographic scanning; such reciprocity properties have been applied elsewhere and may serve almost any optical scanner as long as the acceptance aperture subtends sufficient backscatter to collect a signal that overrides systematic noise. It is noteworthy that the patent document (Pol 2) limits use of this hollow cylindrical scanner to include retrocollection, interpreted in Section A1.4. Holographic facets do provide the advantage of spectral selection to allow concentration of the desired signal component upon the detector, a function that would otherwise be provided by auxiliary filtering.

Substantive attention was devoted (Pol 1, Pol 3) to the resolution characteristics of this scanner. As developed in the general discussion of resolution in Section 2.8, this cylindrical implementation is represented within the class of scanners having an aperture d (underilluminated in this case) which is displaced from its rotating axis by a distance r and focusing to an image point over a distance f, yielding the positive augmentation term r/f. Because of its radial symmetry as represented in Fig. 2.8a, the magnification m is unity ($\Theta = \Phi$) and its resolution is expressed by Eq. (18) of Section 2.8:

$$N = \frac{\Theta d}{a\lambda} \left(1 + \frac{r}{f} \right) \tag{1}$$

where Θ is the active scan angle. Another form more similar to those represented in the cited work is expressed by Eq. (16) of Section 2.8:

$$N = \frac{d}{a\lambda} (\Theta + \alpha) \tag{2}$$

where α is interpreted as the full angle converging to the image point P. To unify with the referenced work, aside from some direct nomenclature matching, the d in Eqs. (1) and (2) is expressed there by the paraxial approximation over r, subtended by its small angle $\Delta\Phi_A$ as $d = r\Delta\Phi_A$, where Φ_A is the full aperture (mechanical scan) angle. Their additional duty cycle term $1 - (\Delta\Phi_A/\Phi_A) = (1 - \kappa)$ is expressed here by Eq. (1c) of Section 2.8 as $(1 - \Delta)$.

The form of Eq. (1) is preferred, for it denotes explicitly the fundamental effective change in aperture size when it is displaced by r and converges flux through f (see Fig. 2.5 and its discussion). Thus it correctly yields an unaugmented resolution when $r = 0$. Equation (2) and the corresponding referenced equations (Pol 1, Pol 3) require implicitly that $\alpha = 0$ when $r = 0$. This is confirmed in Eq. (17a) of Section 2.8, where

$$\alpha = \Theta(r/f) \tag{3}$$

Also in the first work (Pol 1) is a comparison with a polygon mirror scanner having the same number of facets, the same radius, the same beamwidth at the facet, and the same duty cycle, concluding resolution differences between the two. Having assumed, however, a prismatic polygon illuminated radially unsymmetric such that the change in reflected beam angle is twice that of the mechanical one [corresponding to $m = 2$ per Eqs. (28a) and (28b) of Section 2.8], there appears a factor 2 in their Eq. (4) for polygon resolution. Had a radially symmetric pyramidal polygon (Bei 6) been selected (for which $m = 1$, as in their radially symmetric holographic scanner), that difference would have disappeared. Also, having assumed that the reflected flux (from the polygon mirror) was collimated rather than convergent ($f = \infty$, for subsequent field flattening; also an option for their holographic scanner), the augmenting factor corresponding to α fails to appear in their Eq. (4) for polygon resolution. Thus, when both scanners provide the same optical functions (in addition to exhibiting the same number of facets, size, and duty cycle), their resolutions are identical. It is also worthy of reemphasis that the option of $m \approx 2$ exists for holographic scanners as well, allowing resolutions corresponding to those of prismatic polygons when illuminated such that $m = 2$.

Since this scanner was formed to be radially symmetric, it provides an ideally circular scan locus in a plane normal to the axis, as do all such configurations, conventional or holographic (see Sections 2.4 and 4.1.5). It was tested over a flat surface, utilizing a sufficiently high output f-number and narrow scan angle to allow tolerable compromise defocusing at the center and margins of scan when the holographic facets are recorded and reconstructed at the same wavelength, 488 nm. For reconstruction at 633 nm, the exposing interfering angles were modified to approach fringe matching at the two wavelengths. The resulting focal spot was tested to maintain reasonable uniformity when measured along the field curvature. No report was provided regarding spot quality across a flat surface.

An attempt to compensate for this aberration and defocus due to both wavelength shift and flat-field scan was conducted elsewhere (Rim) to develop lenses such that their utilization during holographic exposure at 488 nm would allow reconstruction at 633 nm over a flat image surface, seeking compensation at the same time for the tangent (nonlinearity) error over the flat field. Although it is impossible to impart into a single lenticular holographic member perfect compensation for both nonlinearity and the family of aberrations, reasonable compromise was achieved. Elemental resolution was somewhat under 2000 at FWHM (see Section 2.8.1), over a narrow scan angle of $\pm 8°$.

An earlier effort toward linearization on a flat surface (Whi) was to depart from radial symmetry in both formation and reillumination of the holograms (with no wavelength shift). For a drum 100 mm in diameter, optimal linearization was realized when the exposures were 15 mm off center and the reconstruction point was effectively 65 mm off center. For an approximately $\frac{1}{2}$-rad scan angle from an aperture of 2.24 mm (limited by a stop), the reported estimated resolution was 790 points, significantly below the diffraction-limited calculation of about 1500 for those parameters. One wonders, therefore, if simply reducing the scan angle of a radially symmetric system and matching the best flat image (as was described earlier) would not serve equally well. A further discussion of wavelength shift and linearization factors appears in Chapter 5.

4.2.1.3.3. Transmissive Conic Substrate Scanners

The conic substrate scanner operates in a manner similar to that of the spherical and cylindrical forms described above, also providing the options for radial symmetry and an output beam normal to the rotating axis. The transmissive configuration is introduced in Section 4.1.6 and the reflective one in Section 4.1.8.4. Two transmissive implementations appear in the literature: one by Kramer (Kra 3) as a variation of his flat disk configuration (subsequently discussed in Section 4.2.2.4) and another by Funato (Fun 1) analyzing additional consequences of beam misplacement, discussed in Section 5.3.1. Although not a fundamental constraint, both consider that the diffracted output beam converges to a focus—where, indeed, the identified problem of focal spot shift is aggravated. If operated with a collimated diffracted output, as represented in Fig. 4.22 for the spherical surface (employing subsequent lenticular focusing), the added factor due to beam translation is nulled, leaving only the angular error.

The configuration by Kramer, appearing as Fig. 5 in his patent (Kra 3), is similar to that represented here as Fig. 4.5. The diffracted output ray group converges to a point (focused) on a surface which is assumed cylindrical and concentric with the axis, to satisfy radial symmetry. While extensive discussion is provided for the sensitivity of the output beam angle to tipping of a flat disk-like substrate (analyzed in Section 4.2.2.4), no coverage of this factor appears for the conic substrate. This distinction, along with the convergent output condition, was addressed by Funato in his patent (Fun 1) and clarified in a private communication (Fun 2). The development of these factors appears in Section 5.3.1, where we discuss substrate angular errors. It is noteworthy that the transmissive conic or spherical substrate scanners, when operated in radial symmetry and near the Bragg condition, can provide perfectly straight line scans while achieving significant insensitivity to substrate wobble.

4.2.2. Scanners Having No Surface Component Normal to the Rotating Axis

4.2.2.1. Transmissive Disk-like Scanners

The first published work on this form of scanner was by Cindrich in 1967 (Cin), providing valuable consideration of its characteristics; primarily in disk form when illuminated with coaxial collimated light, as represented in Fig. 4.3*a*. He showed that satisfaction of radial symmetry in holographic reconstruction provides an output that maintains radial symmetry (although nonnormal to the axis) and a stationary image quality during hologram rotation. Also, that operation with off-axis reillumination (i.e., nonradial symmetry) may, under special conditions, be utilized to advantage. These characteristics are expressed in Section 4.1 in terms of the CRS (conformal ray surface) for a variety of substrate configurations and expanded for departure from symmetry in Chapter 5. Cindrich also reported the results of experiments that confirmed some of the anticipated results. A hologram was exposed with a collimated reference wave and a spherical object wave derived from a spatially filtered focal point. When reilluminated with the conjugate to the reference wave, a (near) diffraction-limited point (as determined by an approximately $f/7$ converging cone at $\lambda = 633$ nm) was reconstructed. (See Section 2.8.1 for spot size and resolution considerations.) Also expressed in that work is the option of providing an array of overlapping holograms on a small disk substrate, or enlarging it to accommodate adjacent nonoverlapping ones, more typical of those currently utilized. Such selection, relating to multiplexing, is discussed in Section 4.4.

Since the transmissive disk was historically one of the first types of holographic scanners investigated (see Appendix 1), it was long appreciated that unless something was done to allow its generation of a flat and linear scan, it would be destined to limited service scanning arcuate functions. The same is true of the reflective disk-like devices, which saw high early interest in scanning circular arcs. We shall now review the reflective systems and then return to the transmissive ones when discussing the rectification of their arcuate function into straightened lines.

4.2.2.2. Reflective Disk-like Scanners

The earliest reflective disk-like holographic scanner was the concave configuration developed by the author and reported along with the spherical (convex)
Holofacet and flat disk reflective scanners in 1973 (Bei 4). It is reviewed in the
forthcoming section and in Appendix 1. The first (3-in.-diameter) research
model of that scanner, designed in early 1969, is on loan at Boston's Museum
of Science. A reflective flat disk scanner is illustrated in Fig. 4.24 utilizing
collimated reference and (overilluminated) reconstruction beams. Options are
available in both for underillumination and nonfocusing outputs. The concave
disk-like scanner merits attention because of its unique wavelength-shift accommodation, derived from the Rowland circle properties of the concave grating.

4.2.2.2.1. Concave Holofacet Scanners

The concave Holofacet scanner is illustrated in Fig. 4.25. The concave surface
is spherical and centered on point C, which corresponds to the focal point of
the reference beam during holographic formation and the reconstruction beam
during reillumination. The exposure and processing procedures are similar to
those detailed for the convex spherical system described in Section 4.2.1.1.
While the reconstruction beam is shown overilluminated, this is as tested; underillumination is a clear option. Stigmatic wavelength shift requires that the
output be focused, although a long focal length may be designed to operate
through a flat-field lens. The output beam is drawn from the oppositely disposed
Holofacet to provide a more favorable cross-scan aperture from the curved substrate.

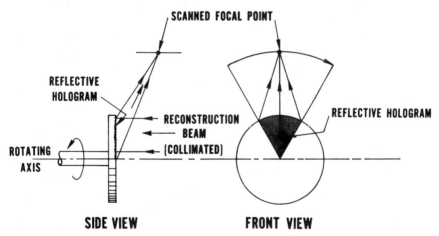

Figure 4.24. Reflective flat disk scanner employing collimated reconstruction beam. Over- or underillumination and nonfocusing outputs are free options.

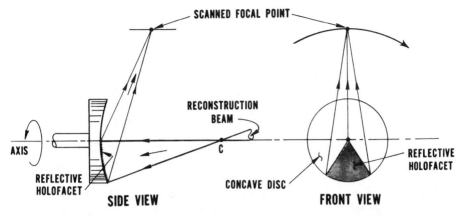

Figure 4.25. Reflective concave Holofacet scanner, overilluminated. Reconstruction beam focuses through spherical center C to illuminate hologram which diffracts output to scanned focal point. System provides unique wavelength shift accommodation (Figure 4.46).

It is useful to consider the along-scan resolution of this configuration in light of this oppositely derived output beam. Following the development in Section 2.8.2, we note that in this radially symmetric case, the optical angular change Θ is equal to the mechanical one Φ (for unity magnification), wherein the resolution is expressed with the augmentation term $(1 + r/f)$. Reviewing Fig. 2.5 in that section, we note that when r and f are both on the same side of the origin, they are taken as positive, and augmentation is additive. If, however, f is here considered positive, then r need be negative, and the augmentation actually *reduces* resolution by the factor $(1 - r/f)$. We see, therefore, that in this configuration a small sacrifice in along-scan resolution provides a significant increase in cross-scan aperture (for a given output beam inclination to the axis).

The wavelength-shift accommodation of this concave grating configuration represented a major motivation for its development into a scanner. While wavelength shift and its control are covered in Section 5.4, discussion of the unique characteristics of this system is appropriate here. The problem is the aberration suffered by a reconstructed wave when a nonplanar and nonlinear (holographic) grating is reilluminated by a wavefront whose image wavelength λ_i differs from its object exposure wavelength λ_o. As described in Section 4.2.1.1, this relief reflective grating is developed from exposed photoresist which exhibits principal sensitivity in the wavelength region below 450 nm. Yet it is very desirable to make the system operative at the He-Ne wavelength of 633 nm and at longer wavelengths provided efficiently by laser diodes.

Figure 4.26 illustrates the wavelength shift condition developed by Jobin-

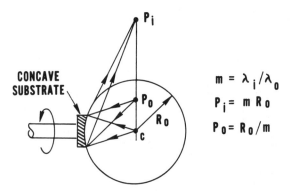

Figure 4.26. Concave grating conditions for wavelength-shift accommodation. P_i is distance CP_i at image wavelength λ_i; P_o is distance CP_o at object wavelength λ_o.

Yvon and described in the first publications (Bei 4, Fla 1). The concave substrate having a radius of curvature R_o is centered at C on the rotating axis. Upon selecting the desired image point location P_i (measured from C), a straight line connecting P_i to C intersects object exposure position P_o (measured to C) such that

$$P_i = mR_o \quad \text{and} \quad P_o = R_o/m \qquad (4a,b)$$

in which

$$m = \lambda_i/\lambda_o \qquad (5)$$

where subscript i denotes image reillumination wavelength and o denotes object exposure wavelength. The image point P_i may be located anywhere outside R_o (and not necessarily on a normal to the axis illustrated as a special case in Fig. 4.26) and is reconstructed stigmatically following these Rowland circle harmonic conjugate relationships. Equations (4a) and (4b) also yield directly

$$P_i/P_o = m^2 \quad \text{and} \quad P_iP_o = R_o^2 \qquad (6a,b)$$

with m called the harmonic constant.

The experimental scanner was designed and tested to provide the extremely high resolution of 50,000 elements/scan over a 25-in. circularly curved format ($\frac{1}{2}$-mil spot size). Three Holofacets were deposited on the spinner of radius 2.25 in. to provide a scan radius of 12 in. The holograms were exposed at 488 nm (argon-ion) and reconstructed stigmatically at 633 nm (He-Ne) wavelengths. The shaft speed and bandwidth were 2400 rpm and 5 megapixels/sec, respectively. The earlier (3-in.-diameter) research model is discussed further in Appendix 1, Section 1.3.

4.2.2.2.2 Reflective Flat Disk Scanners

During the early 1970s, John W. Locke of the University of Toronto initiated one of the most demanding development efforts (Loc), intending to record NASA ERTS and Landsat images in full color using a rather complex assembly of holographic scanner and 9-in. continuous (cupped) film transports. A historical review of this heroic work is represented in Appendix 1, Section 1.3. The approach at first observation appears similar to that of the flat reflective disk introduced in Section 4.1.8.5. Some differences exist, however, which merit attention.

Referring to Fig. 4.27, a reproduction from the Locke 1974 patent (Loc), the system may be described as an overilluminated reflective disk scanner,* radially symmetric and providing a circularly focused locus upon correspondingly curved recording medium. Although one film transport is shown in this figure, the system, designed for the Canada Centre for Remote Sensing (CCRS) in Ottawa, was designed to expose *four* sets of images on four film media whose transports were disposed symmetrically about the scanner.

This reflective configuration and that described in Section 4.1.8.5 differ in the method of holographic exposure and utilization. Whereas the reference wave shown in Fig. 4.9 converges to a point P_o beyond the disk and the reconstruction wave is formed as its conjugate from p'_o, the reference and reconstruction waves utilized by Locke were both derived from the same point (through spatial filter 12 of Fig. 4.27). Since the reference and reconstruction waves are not mutual conjugates, the object and reconstructed image are also not mutual conjugates, and the object beam in Fig. 4.27 must be incident on the disk converging to the virtual point B. The reconstructed image at A is thus a reflected continuation of the object beam. This operation is analogous to a conventional transmission hologram with an added reflecting surface.

When the disk is exposed with a collimated wave normal to its plane surface, the valid reconstruction beam is identical to the reference beam (Fig. 4.24), also collimated normal to the disk. The reconstructed point is then the same point from which the object wave was directed in hologram formation. The same situation exists in Fig. 4.25, for which the reference and reconstruction beams are identical as self-conjugates, and the reconstructed wave retraces conjugately to the original object wave. This process of reconstruction to the same point as used in construction (when reilluminated identically) can be important

---

*The hologram was formed on a silver halide plate. Aluminizing the emulsion side after photoprocessing forms a reflective grating of the residual relief surface (Rig).

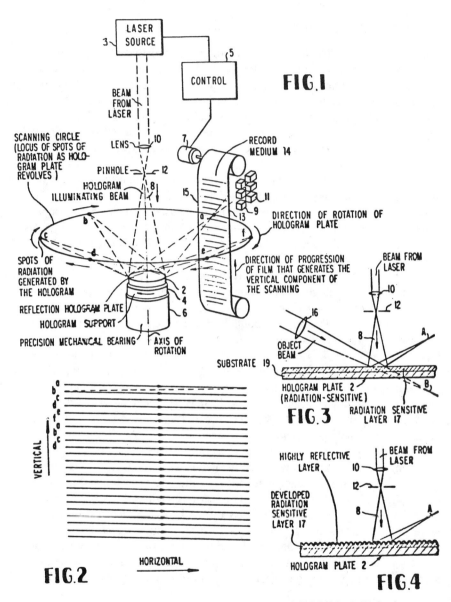

Figure 4.27. Reproduction of illustrations in U.S. Patent 3,795,768 by J. W. Locke.

for the practical need for uniformity of spot relocation and quality from the multiplicity of holographic facets on the scanner.

Another noteworthy factor is that the multicolor image points were to be derived from overlapping gratings on the same scanner surface. Implications regarding this choice are covered in Section 4.4.1.

4.2.2.3. Disk-like Scanners with Auxiliary (Axiconal) Optics

All the disk-like scanners discussed thus far exhibit their characteristic output beam projection which is nonnormal to the rotating axis. Although this is not overly burdensome for those systems having concentric image-bearing surfaces (such as the surface $S(r)$ in Fig. 4.1 and that of Locke just discussed), those that require flat-field scan are not accommodated by the skewed (conically shaped) output projection unless utilized as arcuate scans on flat surface $S(x, y, Z_1)$ per Fig. 4.1. These options are seldom utilized. Flat-field scans are most often (effectively) straight-line scans. There are also times when even the arcuate scan into a concentric cylinder needs to be normal to the axis—such as for retroreflection from normal incidence of the focused beam upon the local image surface—or when the output skew angle is so extreme as to enlarge the spot excessively in the cross-scan direction. For these reasons it is desirable to be able to transform the skewed output beam of a disk-like scanner to normalcy to the rotating axis.

A method for accomplishing this with the addition of a stationary optical component was developed by C. Ih (Ih 1, Ih 2). Published in the 1976–1977 period, it was introduced substantially earlier by Ih as a variation of the reflective disk Holofacet scanner, reported by L. Beiser at a conference in 1974 and published in its proceedings in 1975 (Bei 7). Appendix 1 provides more historical background. The technique eliminates some constraint on the shape of the scanner and is applicable to transmissive as well as to reflective systems. As illustrated in Fig. 4.28a, the added component is a section of a convex spherical auxiliary reflector (AR) mounted such that it reflects the skewed principal output ray normal to the axis. Although the AR is not limited to a spherical surface shape, this form is preferred because of its ease of accurate fabrication. Any radially symmetric reflector or refractor (axicon) may, in principle, serve this function. It may even be a diffractor.

To complement the aberration that would otherwise develop by the added component (which must be curved for radial symmetry), it or its "copy" is employed in the original exposure of the hologram H, such that its object wave is predistorted by the AR on its path to H. Upon reillumination of H, the aberrated reconstructed wave retraces its path conjugately and (if desired and controlled) becomes stigmatic upon reciprocal action by the auxiliary optics (Col).

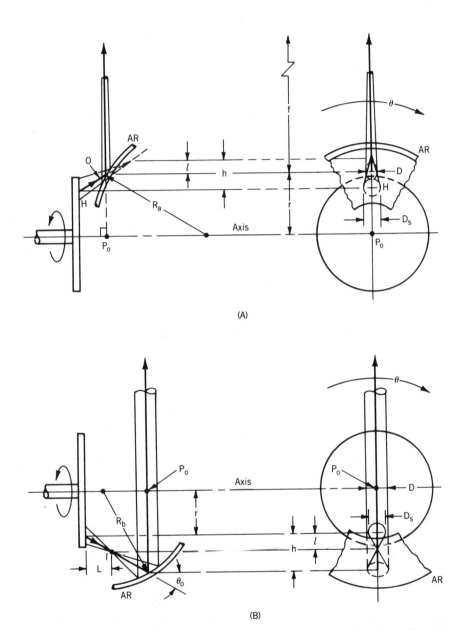

Figure 4.28. Disk scanner and auxiliary reflector (AR). Principal ray O reflected normal to axis throughout scan. (*A*) Convex AR. (*B*) Concave AR.

The auxiliary optics may be concave if situated on the opposite side of the axis, as illustrated in Fig. 4.28*b* to provide an output beam which, after focusing at *f* and reflection from the AR, propagates in the same direction as before. (Ih2, Ih6) This concave case is preferred for directing a collimated output beam into a flat-field lens, for which the pupil relief distance can be arbitrarily small due to close access to the nodal point at p_o. For flat-field application, the convex case (Fig. 4.28*a*) is at a disadvantage, for the lens must be situated sufficiently beyond the AR to avoid mechanical interference. This can increase its pupil relief distance and consequential size and cost.

4.2.2.3.1. Design Parameters

To develop a design approach, Fig. 4.28 is redrawn in Fig. 4.29 for analysis. Consider first (Fig. 4.29*A*) the convex case (above the axis). Hologram *H* diffracts principal ray *O* through the angle Θ_o to be incident on mirror M_a. To be reflected to $O_\perp$ normal to the axis, its local angle to the axis α_a at *o* must be*

$$\alpha_a = \frac{\pi}{4} + \frac{\Theta_o}{2} \qquad (7)$$

If M_a is a straight line in the meridian plane shown, radial symmetry requires that this form a conic reflector with its apex at *a*. For M_a spherical, radial symmetry requires that its radius R_a be centered on the axis at *c*. To determine the magnitude of R_a, observe that *V* (the extension to $O_\perp$) is common to both triangles containing angles α_a and β_a. $V = r_a + h$, where $h = s_a \tan \Theta_o$ and $V = R_a \sin \beta_a$. Thus

$$r_a + s_a \tan \Theta_o = R_a \sin \beta_a \qquad (8)$$

whence

$$R_a = \frac{r_a + s_a \tan \Theta_o}{\sin \beta_a} \qquad (9)$$

Since right triangle *aoc* requires that β_a complement α_a,

$$\beta_a = \frac{\pi}{4} - \frac{\Theta_o}{2} \qquad (10)$$

---

*This applies to all (reflective) auxiliary surfaces, for we consider here only the principal ray. All other rays of the flux bundle are corrected by holographic exposure through M_a (the auxiliary reflector or equivalent).

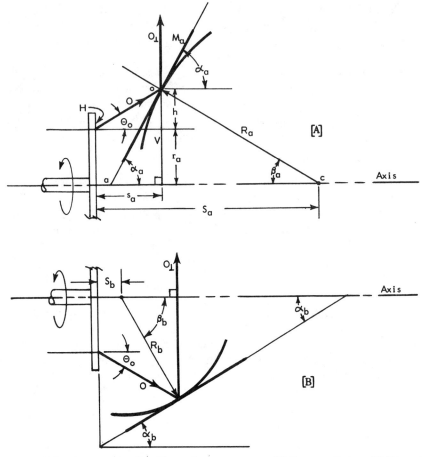

Figure 4.29. Disk scanner and auxiliary reflector parameters. (A) Convex reflectors. (B) Concave reflectors.

and

$$R_a = \frac{r_a + s_a \tan \Theta_o}{\sin (\pi/4 - \Theta_o/2)} \qquad (11)$$

The center of radius R_a is situated at c a distance S_a from H, where

$$S_a = s_a + R_a \cos \beta_a = s_a + \frac{r_a + s_a \tan \Theta_o}{\tan \beta_a} \qquad (12)$$

Following an analysis similar to that for Fig. 4.29B, in which the AR is below the axis, corresponding relationships are derived. Dropping subscripts *a* and *b* (for above and below the axis), the equations are combined with the use of double signs (the upper one representing above the axis; the lower one, below the axis), yielding

$$\alpha = \frac{\pi}{4} \pm \frac{\Theta_o}{2} \qquad \beta = \frac{\pi}{4} \mp \frac{\Theta_o}{2} \qquad (13a,b)$$

$$R = \frac{r + s \tan \Theta_o}{\sin \beta} \qquad (14)$$

$$S = s \pm R \cos \beta \qquad (15)$$

4.2.2.3.2. *Resolution*

The resolution provided by scanners with axiconal auxiliary optics has been subjected to substantive analysis (Ih 2) because of the apparent complexity introduced by the extra components. We provide a simple interpretation of the resolution of such scanners, dramatically bypassing the quantity of variables introduced to account for a variety of ray-trace paths and lenticular powers. In so doing, we reduce drastically the opportunity for errors in analysis and interpretation.

Considering the illustrations in Fig. 4.28 observe that since the auxiliary optics is radially symmetric, *performance is unchanged if it rotates with the holographic disk.* In fact, the auxiliary optics can be considered as the scanner, reducing the problem to that of determining the resolution from the auxiliary optics, where outputs are stigmatic and conveniently normal to the axis—where design equations are simple and rigorous.

The case of collimated output from the concave AR (Fig. 4.28*b*) is considered first. Assume initially that the scanner is overilluminated. We conclude immediately that the nodal point is p_o on the axis. The aperture size D executes an angular scan Θ for which resolution is simply

$$N = \frac{\Theta D}{a\lambda} \qquad (16)$$

by Eq. (4) of Section 2.8.1. Since D is taken at the nodal region, resolution is unchanged if the output beam is uncollimated, that is, converging (to positive focus) or diverging (from negative focus). While the problem reduces to determining the value of D, this is usually easily derived. For the collimated condition, similar triangles yields $D/D_s = (h - l)/l$, where D_s is the scanner aperture width and h and l are identified in the figure. Thus the resolution equation for this case related back to the scanner becomes

$$N_B = \frac{\Theta D_s}{a\lambda} \left(\frac{h}{l} - 1 \right) \qquad (16a)$$

Having shown earlier (Section 2.8.3) that the resolution is invariant with under- or overillumination, the same equation holds for underillumination, with D_s representing the illuminated portion.

Considering now the case shown in Fig. 4.28a, this takes the form of the displaced deflecting aperture, for which the resolution is given as [by Eq. (11) of Section 2.8.2]

$$N_A = \frac{\Theta D}{a\lambda} \left(1 + \frac{r}{f} \right) \qquad (17)$$

The terms r and f are identified in the figure. Clearly, for the output collimated ($f = \infty$), it reverts to the earlier Eq. (16). Again, to relate this to the scanner aperture D_s, we note from the figure that $D = (l/h)D_s$, whence the resolution becomes (using notation of Fig. 4.28a)

$$N_A = \frac{\Theta D_s l}{a\lambda h} \left(1 + \frac{r}{f} \right) \qquad (17a)$$

The underilluminated resolution is similarly unchanged [by Eq. (19) of Section 2.8.3].

4.2.2.3.3. Correction for Scanner Wobble

Historically, this development followed the one discussed in the next section on input–output beam angle relationships. The results of that analysis were combined with an appreciation of the focusing characteristics of the concave AR (Kra 5) to yield a unique solution to the disk wobble problem. Portions of both disciplines are needed to explain the process.

Referring to Fig. 4.28b (side view), let us imagine that the marginal rays shown to focus through f are, in fact, principal rays which are so directed during disk wobble that they converge through f. If that is achieved, the resulting output rays reflected from the AR will be collimated and paraxial. (*Note:* These error components represent very small angles of up to a few minutes of arc— well within the approximation of spherical concave mirror focus characteristics.) The collimated output principal rays, though translated, when focused through a flat-field objective lens, will converge to the same point, nulling scanner wobble. The problem now reduces to establishing the input and output angles and beam positions that make this happen.

Analysis in the next section yields Eq. (29), repeated here:

$$d\Theta_o = \pm \left[1 \mp \frac{\cos (\Theta_i + \alpha)}{\cos (\Theta_o \mp \alpha)} \right] d\alpha \qquad (18)$$

a relationship between the change of output angle $d\Theta_o$ due to changing wobble angle $d\alpha$ as a function of the input and output angles Θ_i and Θ_o. It applies to both transmission and reflection gratings, the upper sign for transmission and the lower for reflection. Also, following the development by Kramer (Kra 5) and adapting to our notation, the output angle from the tilted substrate geometry that satisfies focusing through f is given by*

$$d\Theta = - \frac{r \sin \Theta_o \cos \Theta_o}{L} d\alpha \qquad (19)$$

in which the substrate tilt angle is negative due to the sign convention.

Upon equating Eqs. (18) and (19) there results (for $\Theta_o > \Theta_i$)

$$r = - \left[1 \mp \frac{\cos (\Theta_i + \alpha)}{\cos (\Theta_o \mp \alpha)} \right] \frac{L}{\sin \Theta_o \cos \Theta_o} \qquad (20)$$

*The reader interested in the derivation of Eq. (19) from the original reference (Kra 5) will be aided by the following clarifications, utilizing the notations appearing in that reference: Eq. (2) states that $d\Theta = -dh/r$, for which r is not identified in the associated Fig. 3 and h is not normal to the angle $d\Theta$. Letting r = the distance from the disk to the focal point R in Fig. 3, and the normal distance $\approx h \sin \Theta_d$, then

$$d\Theta = - \frac{dh \sin \Theta_d}{r}$$

Also, with $\cos \Theta_d = H/r$, then

$$d\Theta = - \frac{dh \sin \Theta_d \cos \Theta_d}{H}$$

With $\Phi = h/D$, then $dh = D \, d\Phi$, yielding

$$d\Theta = - \frac{D \sin \Theta_d \cos \Theta_d}{H} d\Phi$$

which is the balance of Eq. (2) in Ref. Kra 5, and the equivalent of our Eq. (19).

With $l/L = \tan \Theta_o$, the second term may also be expressed as $l/\sin^2 \Theta_o$. Since the tilt error α is small (nominally zero) and we seek the magnitude of r, Eq. (20) simplifies to

$$r = \left(1 \mp \frac{\cos \Theta_i}{\cos \Theta_o}\right) \frac{L}{\sin \Theta_o \cos \Theta_o} \tag{21}$$

(upper sign for transmission and lower for reflection).

Although this interesting technique for wobble correction can also render the output beam normal to the axis for effective straight-line scan, it suffers from alignment difficulty and sensitivity to deflector centration. This can be appreciated from the critical orientation of the focal point f with respect to the auxiliary reflector. Further, as typical for focusing-type holographic elements (upon which this system depends), it is not readily adaptable to stigmatic operation under wavelength-shift conditions (between exposure and reconstruction). We investigated one possible solution: adaptation of the concave reflective grating (Fig. 4.26) reversing the roles of P_o and P_i (since one must be inside and the other outside the Rowland circle). For this concave grating case, $\Theta_i = 0$ (illumination normal to the surface) and Eq. (21) reduces to

$$r_c = \left(1 + \frac{1}{\cos \Theta_o}\right) \frac{L}{\sin \Theta_o \cos \Theta_o} \tag{22}$$

Substituting into Eq. (22) for $30° < \Theta_o < 60°$ yields the value of L to range between approximately $\frac{1}{5}$ to $\frac{1}{7}$ that of r_c—unfortunately, a very small focal distance indeed—and not adaptable to wavelength shift operation per Fig. 4.26.

4.2.2.4. Disk Scanners Operating in the Bragg Regime

A significant variation to holographic scanner design is, effectively, operating the gratings in the Bragg regime. This condition, introduced at the end of Section 3.2, occurs when the input and output angles are equal (with respect to grating surface normal), that is, when $O_i = O_o$ about II, by Fig. 4.30A.

The principal consequence is that when the Bragg condition is satisfied, small changes in angle of the holographic disk $\Delta\alpha$ (with respect to a fixed axis) result in a substantially reduced effect on the output angle Θ_o. This alleviates one of the most insidious errors of line placement in the cross-scan direction, due, for example, to wobble of the scanner upon rotation about its axis. This error, which generates nonuniform line structure (to which we sense high visual acuity), is normally difficult to control—if not by (costly) precision of spinner and bearing assemblies, then by the addition of either high-speed beam-position

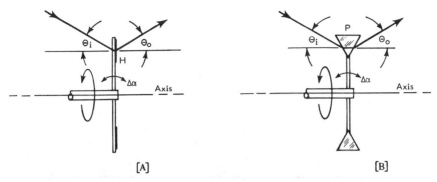

Figure 4.30. (*A*) Holographic scanner in Bragg regime. (*B*) Hypothetical prismatic scanner at minimum deviation.

techniques (which require precalibration or real-time measurement of error), or by anamorphic optics, discussed in Section 5.1.

To appreciate more effectively the contribution of operating a holographic diffractor near the Bragg angle, consider the analogous condition of operating a conventional prism near its angle of minimum deviation. As is well known from classical optics (Jen 2), a prism will divert an incident beam through an angle of minimum deviation when the input and output angles are symmetric about its apex. This condition is illustrated with prism P on a hypothetical scanner in Fig. 4.30*B*. The rate of change of Θ_o with respect to α is zero at its null and increases slowly for small values of $\Delta\alpha$. This analogy will be developed further following analysis of the Bragg null condition.

4.2.2.4.1.   Analysis of the Bragg and Prism Null Conditions

Assume a linear grating on a plane transparent substrate having parallel surfaces. We adapt the procedure due to Kramer (Kra 3), circumventing, however, the complication of differing indices of refraction of air, substrate, and holographic medium at their several lamina thicknesses and surfaces. In a set of plane-parallel interfaces separating media of differing thicknesses and refractive indices (immersed in the same medium—air), the output angles are independent of these differences. The process can then be viewed as occurring at a single surface supporting hologram H, as depicted in Fig. 4.31.

There, hologram H is mounted vertically, normal to axis A (in plane in-and-out of paper). Input beam I is incident on H at an angle Θ_i with respect to the normal A, and the output beam O is diffracted from H at an angle Θ_o with respect to the normal A. An angular error α is imparted to H and A such that they rotate to H' and A', respectively. Θ_i and Θ_o appear thus as Θ_i' and Θ_o' with the addition and subtraction of α, respectively. We seek a relationship between the output angle Θ_o and the error α.

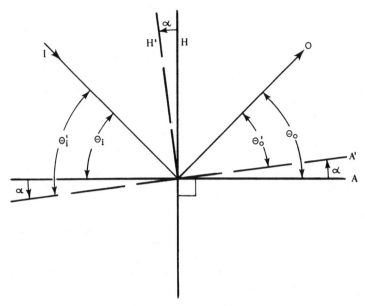

Figure 4.31. Angular and coordinate relationships for analysis of hologram H tipped to H' through angular error α.

The grating equation of a transmission hologram, first order [per Eq. (4) of Section 3.2] is restated for the displaced condition,

$$\sin \Theta_i' + \sin \Theta_o' = \frac{\lambda}{d} \tag{23}$$

whence

$$\sin \Theta_o' = \frac{\lambda}{d} - \sin (\Theta_i + \alpha) \equiv G \tag{24}$$

Differentiating G first with respect to Θ_o' and then with respect to α,

$$\frac{dG}{d\Theta_o'} = \cos \Theta_o' \quad \text{and} \quad \frac{dG}{d\alpha} = -\cos (\Theta_i + \alpha) \tag{25a,b}$$

and dividing Eq. (25b) by Eq. (25a),

$$\frac{d\Theta_o'}{d\alpha} = \frac{\cos (\Theta_i + \alpha)}{\cos \Theta_o'} = -\frac{\cos (\Theta_i + \alpha)}{\cos (\Theta_o - \alpha)} \tag{26}$$

Since

$$d\Theta'_o = d(\Theta_o - \alpha) = d\Theta_o - d\alpha \qquad (27)$$

then

$$d\Theta_o = \left[1 - \frac{\cos (\Theta_i + \alpha)}{\cos (\Theta_o - \alpha)} \right] d\alpha \qquad (28)$$

For a reflection hologram, the sign in Eq. (23) is $(-)$ and $\Theta'_o = \Theta_o + \alpha$ in Eqs. (26) and (27), yielding

$$d\Theta_o = \pm \left[1 \mp \frac{\cos (\Theta_i + \alpha)}{\cos (\Theta_o \mp \alpha)} \right] d\alpha \qquad (29)$$

where the upper sign is used for transmission and the lower for reflection. These expressions apply to any incident/diffracted ray path, and therefore to a collimated ray group diffracted from a plane linear grating. They are expanded in Section 5.3.1 to include the convergent beam displacement error derived from a grating having lenticular power.

The structure of Eq. (29) reveals important consequences. Only in transmission (upper sign) is there a possibility of null. That occurs when the fractional term equals 1, or when $\Theta_i = \Theta_o$ in the vicinity of small α. In reflection (lower sign), there can never be a null, for under all conditions $|d\Theta_o/d\alpha| > 1$. When $\Theta_i = \Theta_o$, $|d\Theta_o/d\alpha| = 2$, as in mirror reflection.

Equation (29) was evaluated theoretically and experimentally for two cases of transmission hologram formation and orientation (Kra 3), providing the data represented in Fig. 4.32. The case of Bragg compliance, selected for $\Theta_i = \Theta_o = 45°$, demonstrates clearly the suppression of output angle misorientation $\Delta\Theta_o$ as a function of wobble angle change $\Delta\alpha$. At $\pm 0.1°$ wobble (a significant error of 360 arc seconds), there appears a residual error of 1.3 arc seconds. At $\pm 1°$, the suppression is about 30-fold, and for the (gigantic) error of $\pm 5°$, the suppression is about 10-fold. The case of noncompliance with Bragg is given for $\Theta_i = 0°$ and $\Theta_o = 60°$, where we see an almost linear relationship between error and its effect. For very small errors, it is linear, where solving Eq. (29) for $\Theta_i = 0$ and $\Theta_o = 60°$ yields $d\Theta_o/d\alpha = -1$ around $\alpha = 0$.

Returning to the analogy of the minimum deviation prism introduced in the preceding section, one finds that (Jen 2) for a 60° apex angle prism, the error characteristic is almost identical in the vicinity of the plotted data of the Bragg condition in Fig. 4.32. For a prism having a refractive index of $n = 1.53$ (operating in air) the minimum deviation is about 40°, corresponding to $\Theta_i = \Theta_o \approx 20°$. This similarity is no accident. It is instructive to develop the transfer

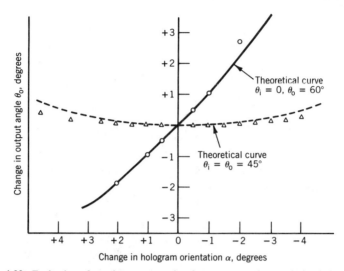

Figure 4.32. Evaluation of angular error equation for two cases of transmission hologram operation. Experimental data (circles for Bragg and triangles for non-Bragg conditions) provide good match with theoretical curves. (After Ref. Kra 3.) See the text for relationship to minimum deviation prism.

equation for the prism and perceive rigorously this interesting consequence of its tipping through the angle α.

Using Fig. 4.30B for reference, let the prism P be composed of material of index of refraction n (immersed in air of $n_o = 1$) and exhibit an apex angle β. Let the input and output angles Θ_{ip} and Θ_{op} be with respect to the face normals, and angles β_i and β_o represent the input and output angles of the propagating ray within the prism, also with respect to the face normals. Expressing Snell's law at both faces,

$$\sin \Theta_{ip} = n \sin \beta_i \quad \text{and} \quad n \sin \beta_o = \sin \Theta_{op} \qquad (30a,b)$$

The relationship between the internal angles and the apex angle can be shown to be

$$\beta_i + \beta_o = \beta \qquad (31)$$

whence, substituting into Eq. (30a), we have

$$\sin \Theta_{ip} = n \sin \beta - n \sin \beta_o \qquad (32)$$

and from Eq. (30*b*),

$$\sin \Theta_{ip} + \sin \Theta_{op} = n \sin \beta \qquad (33)$$

This is recognized as a direct equivalent to the (first-order) transmission grating equation [Eq. (4) in Section 3.2] with $n \sin \beta$ analogous to the grating term λ/d. For β small, λ/d is equivalent to $n\beta$.

Following the prior development for the tipped grating, starting at Eq. (23), we yield the identical result as Eq. (28) for the tipped prism. Although this revelation might motivate consideration of utilizing a prismatic scanner to benefit from its tip insensitivity at minimum deviation, such action will require control of optical integrity and multicomponent uniformity—along with coping with its added inertial and aerodynamic constraints. As in its holographic counterpart, the provision for multielemental optical and mechanical integrity are operational parameters of major practical import, discussed in Chapter 5.

A prism scanner by Rando (Ran) utilizes a prism array in a different manner. Compared to Fig. 4.30*B* the prism is reversed (i.e., apex at the periphery). Instead of utilizing the minimum deviation nulling just described, a double-pass through the prism is stabilized in cross-scan by double reflection of an added (fixed) roof prism, as expressed by Radl (Rad) for reflective scanners. Rando also describes the application of this technique to a holographic disk deflector. As above, similar precautions regarding uniformity within and among each prism or hologram apply, for the double pass will null only complementary errors. The prism applications discussed above are intended for collimated light. For those interested in the use of prisms in convergent light—and contending with the resulting aberrations—the work by Howard (How) is useful and may have implications for the use of linear gratings in convergent light.

4.2.2.5. *Straight-Line Scan from a Plane Linear Grating Deflector*

The Bragg null condition described above inspired one of the most important consequences of operation in that regime—the approximation of a straight-line scan in a nonradial direction (Kra 9). Although a straight scan line can be derived from an output beam launched normal to the rotating axis, the flat disk substrate does not allow normal output launching. As discussed thus far, either the image surface must be correspondingly curved, or auxiliary optics can be employed to redirect the scanning beam normal to the axis. The prospect of deriving a direct straight-line output from a disk-like scanner was so remote that its achievement (albeit an approximation) is here regarded as a milestone innovation. Quite remarkably, a pair of parallel developments (Kra 7, Ant) emerged from disparate locations: the United States and the Soviet Union. Appendix 1 summarizes the progression of events, concluding that these similar developments were independently created. Because of the significance of straight-line generation by a disk-like scanner, both works are detailed further in a corresponding pair of appendices: Appendix 2 is an annotated translation of the published work by Antipin and Kiselev, and Appendix 3 is an annotated abstract from the Kramer patent. Both are augmented with significant interpretation to text and illustration, to fill voids in derivations and to provide continuity of presentation.

Although both perceived the same problem, the motivations and analytic routes were distinctly different. Yet they yielded identical results for a particular set of conditions. Antipin and Kiselev direct their attention to the task of recording real-time video images on motion picture film, while Kramer identifies more with the field of nonimpact printing for business graphics (Kra 6). Figure 4.33 typifies the approach rendered by Kramer and is expanded for analytic development in subsequent discussion. This system is rotationally asymmetric, as represented earlier in Figs. 4.3 and 4.30. In a rotationally symmetric system the input beam angle $\Theta_i = 0$ when collimated. The folding mirror contributes only illustrative and packaging convenience; in no way does it form an auxiliary optic, as does the axiconal AR in Fig. 4.28.

Other methods for deriving straight-line scans (from rotational systems), in-

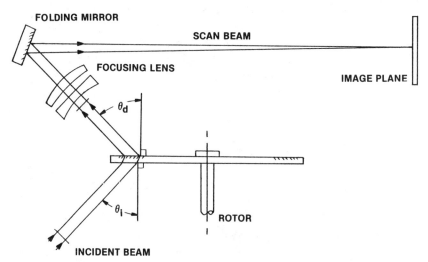

Figure 4.33. Plane grating holographic disk scanner approximating straight-line scan over limited distance on image plane. (After Ref. Kra 6.) © Laser Focus, The Magazine of Electro-Optics Technology, 1981. Reprinted with permission of Penwell Publishing Co.

cluding use of tandem grating action, special lenticular elements, and ana-morphic optics, will be described following discussion of the analytic tech-niques introduced above.

4.2.2.5.1.   Development of Straight-Line Scan

To place the approaches in perspective, we introduce two illustrations; one from each of Appendixes 2 and 3. Figure 4.34 is an illustration from Appendix 2, a transformation of their Fig. 2, with special attention devoted to clarify graphics and remove ambiguity. Figure 4.35 is an illustration from Appendix 3, an ad-aptation of Fig. 5 in the original work, augmented significantly with annotation used in Appendix 3.

Although the nomenclatures in Appendixes 2 and 3 differ from each other as well as from much of that appearing elsewhere in this book, it is judicious to retain here the original notations utilized in the respective discussions, to allow direct connection with Appendixes 2 and 3. The analytic techniques are also fundamentally different: Appendix 2 utilizes a vector approach, while Appendix 3 is primarily geometric. The coordinate systems are, however, similar; the diffractor plane is located in the x-y plane, rotatable about the z axis.

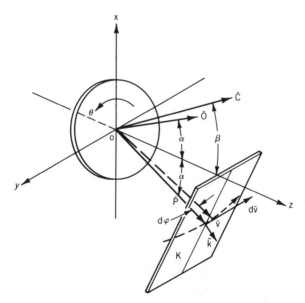

Figure 4.34. Clarified Fig. 2 (Fig. A.2.2′) from Appendix 2. Linear hologram in x-y plane rotates through Θ. Vectors $\overline{C}$, $\overline{O}$, and $\overline{P}$ in x-z plane. $\overline{C}$ is reconstruction wave and $\overline{v}$ is output wave in-cident on image plane K.

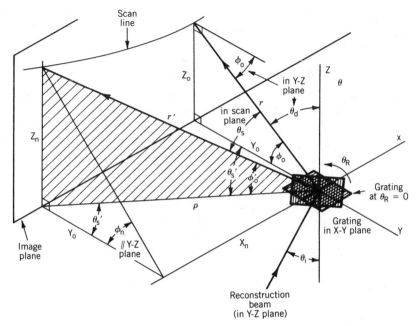

Figure 4.35. Annotated Fig. 5 of Ref. Kra 7 from Appendix 3. Sag in Scan line = $Z_n - Z_o = \Delta Z$.

The problem can be expressed using Fig. 4.35 by observing that the bow in question is represented by the value of ΔZ in the image plane, where

$$\Delta Z = Z_n - Z_o \tag{34a}$$

$$= r' \sin \Phi'_o - r \sin \Phi_o \tag{34b}$$

For the condition of $\Phi'_o = \Phi_o$ (i.e., $\Theta_i = 0$, so that the system is radially symmetric and Θ_d = constant during the rotation Θ_R), then

$$\Delta Z = (r' - r) \sin \Phi_o \tag{35a}$$

$$= (r' - r) \cos \Theta_d \tag{35b}$$

This reveals the two possible conditions for $\Delta Z = 0$ in a radially symmetric system:

1. Either $r' = r$, which makes the image surface cylindrical and concentric on axis z (defeating the purpose of this exercise), or

2. $\cos \Theta_d = 0$, which requires normal propagation of the output from a disk-like spinner (not feasible without the diffractive surface on the spinner exhibiting a normal component). Both conditions are untenable as the problem is defined.

To approach line straightness, impose the conditions that $\Phi'_o = \Phi_o$ at $\Theta_R = 0$, and that Φ'_o *decrease* with increased Θ_R, such that Φ_n = constant as a function of Θ'_s. (Θ'_s is the scanned angle projected upon the X-Y plane.) This requires that Θ_d *increase* with increased Θ'_s (or Θ_R), achieved if Θ_i (the incident beam angle) *decreases* correspondingly, as Θ_d and Θ_i are related through the grating equation. Quite fortuitously, Θ_i *does effectively decrease* with increased Θ_R. That is, Θ_i projected on the Y-Z plane per Fig. A3.7 and Eq. (5) of Appendix 3 is represented by

$$\tan \Theta_i = \tan \Theta_{io} \cos \Theta_R \qquad (36)$$

in which Θ_{io} is the value of Θ_i at $\Theta_R = 0$. Thus, effective Θ_i ranges from a finite Θ_{io} to 0 as Θ_R ranges from 0 to 90°. Transformation of this perception to the conclusion,

$$\sin \Theta_i = \frac{\lambda}{d} - \frac{d}{\lambda} \quad \text{and} \quad \sin \Theta_d = \frac{d}{\lambda} \qquad (37a,b)$$

requires detailed attention to the annotated development in Appendix 3.

Following representation of the diffraction and scan dynamics in vector form, analysis in Appendix 2 develops the change of the output angle φ with rotation Θ (see Fig. 4.34), given by Eq. (9) of Appendix 2 in that notation, for equal angles α between object and reference waves:

$$\frac{d\varphi}{d\Theta} = \sin \alpha \left[4 + \frac{\sin^2 \beta \sin^2 \Theta}{\cos^2 \beta + 4 \sin \alpha \sin \beta \cos \Theta - 4 \sin^2 \alpha} \right]^{1/2} \qquad (38)$$

when $\beta = 0$ (reconstruction beam coaxial with z),

$$\frac{d\varphi}{d\Theta} = 2 \sin \alpha = \text{constant} \qquad (39)$$

denoting scan linearity. (Notations φ and Θ are reversed in Appendix 2 compared to their use in this book.)

When the reconstruction beam $\overline{C}$ and object beam $\overline{O}$ are coincident ($\alpha = \beta$)

[noting that $\overline{C}$ complementary with $\overline{O}$ rather than with $\overline{P}$ is perfectly acceptable analytically (see Appendix 2)], Eq. (37) reduces to Eq. (9a) of Appendix 2:

$$\frac{d\varphi}{d\Theta} = \sin \alpha \ (4 + \tan^2 \alpha \sin^2 \Theta)^{1/2} \tag{40}$$

When $\alpha = 45°$,

$$\frac{d\varphi}{d\Theta}(\alpha = 45°) = \frac{\sqrt{2}}{2}(4 + \sin^2 \Theta)^{1/2} \tag{41a}$$

$$= c \sin \alpha \tag{41b}$$

where $c = (4 + \sin^2 \Theta)^{1/2}$ and is clearly more constant as $\sin^2 \Theta \ll 4$ (small scan angle). For $\sin^2 \Theta \ll 4$, $d\varphi/d\Theta \approx 2 \sin \alpha$, as for $\beta = 0$ per Eq. (39), whereupon $d\varphi/d\Theta = \sqrt{2}$ for $\alpha = 45°$, as concluded also in Appendix 3. $d\varphi/d\Theta$ may be recognized as the parameter magnification m, developed in Section 2.8.4, with particular note regarding $m = \sqrt{2}$ near center of scan for this type of scanner, appearing at the close of that section. Scan magnification and nonlinearity are developed further in Section 5.2.

Appendix 2 continues with a determination of minimum line bow, based on the minimization of $dz/d\Theta$, the change of the intersection point of $\overline{v}$ on plane K (see Fig. 4.34) in the z direction, as a function of hologram rotation Θ. Since it was shown earlier that $d\varphi/d\Theta \approx$ constant, and we seek a minimum, it is unnecessary to complicate this quest with an explicit representation of the bow function on the image plane. Equation (12) of Appendix 2 then yields the condition for minimization as

$$v_x v_z' - v_z v_x' = 0 \tag{42}$$

where the unprimed v's are the direction cosines of the scanning vector $\overline{v}$, and the primed v's are their derivatives with respect to Θ.

Following significant mathematical operation annotated in Appendix 2, this yields the key relationship of Eq. (14) in that work,

$$\sin \beta = \frac{1}{\cos \Theta}\left(2 \sin \alpha - \frac{1}{2 \sin \alpha}\right) \tag{43}$$

$$= \frac{1}{\cos \Theta}\left(\frac{\lambda^2 - d^2}{\lambda d}\right) \tag{43a}$$

which is represented as Eq. (18*a*) of Appendix 3 and appears in this section approximated as Eq. (37*a*).

4.2.2.5.2. *Straight-Line Scan from Counterrotating Plane Gratings*

This form of straight-line scan was proposed (Bra 2, Wya) long before the method described in Section 4.2.2.5.1. (For a chronological perspective, see Appendix 1.) The precept is analogous to the action of synchronously counter-rotating wedges (Lev), utilizing a second wedge in tandem to null the bow component generated by the first one.

Consider a single plane linear grating illuminated normally ($\Theta_i = 0$). If rotated in Φ, the output beam would generate a conic scan locus about the axis ($\sin \Theta_o = \lambda/d = $ constant). Follow this grating with an identical one parallel to the first, with its rotational phase such that the grating lines are parallel at $\Phi = 0$ and the diffracted first orders are complementary, whereupon the output beam will be restored to paraxial at $\Phi = 0$. Now, counterrotate both gratings 90°. The grating lines will again be in phase; this time the first orders will be additive, doubling deflection (for small angles). Note the plane established by this new output angle and the axis. When both gratings are counterrotated synchronously, the output components transverse to this plane are nulled, leaving only the double component which resides in this plane of symmetry, in the direction of $\sin \Phi$. Thus the new output angle Θ_o appears as (Wya)

$$\sin \Theta_o = \frac{2\lambda}{d} \sin \Phi \qquad (44)$$

Wyant continues with a proposal to tend to linearize this function by assuming that the (collimated) output beam is imaged through a distortion-free flat-field lens to a focal point on a flat surface, such a lens having a transfer function $l = f \tan \Theta$ ($l = $ displacement on image plane and $f = $ focal length). Then, as Θ increases, this provides a tangent expansion to the sinusoidal compression, resulting in

$$l = f \tan \left[\sin^{-1} \left(\frac{2\lambda}{d} \sin \Phi \right) \right] \qquad (45)$$

and a demonstrable improvement in linearity (as a function of λ/d) over a limited scan range. Clearly, this technique of lens matching, discussed further in Section 5.2.2.1, may be applied to any such nonlinear scan function that tends to compress as Θ increases.

Nonlinearity notwithstanding, a useful concept appears expanded from earlier work (Loh 2), whereby rigorous synchronism is established rather automatically. By illuminating the "lower" portion of a disk with the paraxial input beam and then folding the first-diffracted output back with a roof reflector and propagating the beam through the "upper" portion of the disk, this provides effective counterrotation from opposite sides of the same disk. Not identified, however, is that because of the extended spacing between disk portions, the beam translates significantly during the second pass, and both the disk and subsequent focusing optics must be large enough in aperture to accommodate this beam wander.

Another variation was proposed which falls into the category of multiplexing (Section 4.4), whereby the disk is composed of two sets of gratings, one at a smaller radius than the other. With proper beam programming, four recurrent scans per disk rotation can be provided with up to 100% duty cycle. This is allowed because the gratings can be overlapping during the scan "flyback" times.

This general technique of counterrotation of plane linear gratings is not known to have been tested operationally. Expressed concerns relate to the purity of the diffractive orders. While blazing or thick holographic techniques can rightfully be applied to the first grating, for $\Theta_{i1} = 0$ and $\Theta_{o1} = $ constant, the second grating sees Θ_{o1} as a varying Θ_{i2}, complicating the output with diffractive artifacts and varying throughput efficiency. It is possible to optimize the angles for maximal purity and uniformity over the limited scan range which provides adequate linearity. This system exhibits a clever combination of attributes and, although long passed its inception, merits a fair appraisal.

At the close of Section 4.4.3.1 is expressed an interesting method which is effectively counterrotation of lenticular gratings in a transmissive drum configuration. The incident beam traverses the drum twice, to form an output scan in which the aberration developed on the first pass is complemented on the second one.

4.2.2.5.3. Straight-Line Scan with Bow Correction

The initial development of Bragg operation yielded $\Theta_i = \Theta_o = 45°$ for both wobble minimization and straight line scan, typified by the geometry of Fig. 4.33. Variations are instituted, for example, by IBM ($\theta_i \approx 22°$, $\theta_o \approx 45°$) for application to P.O.S. scanning (Section 4.2.2.6.1.1), where some line bow is tolerated. One motivation to operate at shallower angles is the relaxation in forming fine pitched gratings. That is, when $\Theta_i = \Theta_o = 45°$, the grating spacing requirement is $d = 0.707 \lambda$ (per Section 3.2), imposing extreme demand upon its sustenance when λ is in the visible region. When $\Theta_i = \Theta_o = 30°$, for example, the Bragg condition is equally satisifed for wobble reduction, while the grating spacing ($d = \lambda$) is relaxed by $\sqrt{2}$. Residual bow is, however, accentuated. To complement this bow, two general approaches may be utilized, individually or jointly:

1. If the bow is small enough, the classic technique of dynamic correction, such as by acoustooptic deflection can be instituted. This provides relatively low resolution high speed deflection for programmed correction during the scanned line. It is, however, relatively costly, unless acoustooptics is already planned for intensity modulation and low resolution deflection can be accommodated by moderate added control over the acoustic drive frequency.

2. Optical compensation may be utilized, such as by anamorphic action described in Section 5.1.2.1 and employed by Datagraphics, Inc. in their microimage recorder. If only bow is to be compensated (wobble reduced by other means), then a prism may be disposed in the path of the scanned output beam such that the bow it generates complements that of the scanner over its operating range. This technique is employed in the StraightScan-2P deflector unit by Holotek Ltd. (Kra 10).

4.2.2.6. Disk Scanners Providing Convergent Beam Output

Aside from the reflective scanners that imaged on curved surfaces discussed earlier, the recently covered disk configurations were all designed to scan collimated light which is focused through subsequent (flat-field) optics into an approximated straight line. Although nothing prevents illuminating these devices in such a manner that they scan convergent light, quite clearly the scanned spot quality on a flat surfaces needs to be compromised. Analysis is compounded by perturbations of wavelength shift (per Section 5.4), whereas the plane linear grating is nonaberrating in collimated light. Approaching such analytic consideration, reviewed here, some investigators have explored lenticular holographic configurations to provide a scanned output focused directly on a flat surface.

A general consequence of lenticular focusing is the need for additional control over beam misplacement, for beam output position is now also determined by linear displacement of the lenticule. Thus the freedom gained in eliminating the subsequent focusing optics is complemented by the need for more critical transverse alignment—disk centration errors. Some of the investigators reviewed here report directly on this effect, whereas others do not, depending on intended application. Also, some accept an arcuate scan while others seek to straighten it, depending upon application.

4.2.2.6.1. Convergent Output with Modified Illumination: Arcuate Scan

A series of publications by Ikeda and his associates (Ike 1, Ike 2, Ike 3) documents dedicated activity toward utilization of the relatively simple interferometric zone plate (IZP) for point-of-sale (POS) scanning, in a form that allows magnified scan length and provides some correction for its consequential aberration. The first such effort is represented in a patent filed in June 1976 (Ike 1) which, in addition to discussing several innovative multiple scan configurations described in Section 4.4.3, introduces the option of operating the IZP in a nonconventional manner.

As expanded in Section 3.4, providing a directly related reference on the zone lens, the IZP is formed by the interference of a collimated reference wave with a diverging (or converging) spherical wave whose principal ray is paraxial with the collimated wave and normal to the hologram plane. This forms a set of n concentric fringes of radius r such that the nth fringe radius is

$$r_n = \left[2nf\lambda + (n\lambda)^2\right]^{1/2} \tag{46}$$

and the grating frequency at radius r and focal length f is

$$\nu(r) = \frac{1}{d(r)} = \frac{1}{\lambda}\frac{r}{(r^2 + f^2)^{1/2}} \tag{47}$$

where $d(r)$ is the grating spacing at radius r.

In Section 2.5 and Fig. 2.2c we described a classical IZP utilization in which marginal segments are oriented about the periphery of a disk and illuminated by a collimated beam parallel to the rotating axis (normal to the disk). The locus of the diffracted scanned focal point is then an arc describing the trajectory of the center of the zone lens. To increase this scan length, a magnified implementation is introduced, in which the same zone lens is used in conjugate imaging—that is, the illuminating source moves from infinity (collimated) to point a, which is somewhat more distant from the hologram than its focal length f. The holographic lens then forms a conjugate image point at a distance b such that the thin lens formula $1/a + 1/b = 1/f$ is sustained. Thus the infinity-conjugate IZP is operated as a finite-conjugate lens. The notation IZP, unless otherwise identified, is regarded as the infinity-conjugate form (Sections 3.4.1 and 3.4.3). Defining magnification $M = b(o)/f$, where $b(o)$ is the image point distance from the lens at the center of scan, these conjugate image points are located at $a = [M/(M-1)]f$ and $b(o) = (M-1)a$. As a result of this magnified operation, the image point is now significantly extended beyond its original position (at f) and executes a correspondingly larger arc during scan. However, because of the nonreciprocal illumination (diverging from a rather than collimated from ∞), the focal point at b becomes astigmatic. The correction selected in this implementation is the simple expedient of adding anamorphic (cylindrical power) in the illuminating beam path such that the average of the foci along the image locus are most well corrected. No detail is provided regarding the degree of correction over the scan interval.

More extensive discussion of another novel approach to this aberration correction was provided in a subsequent work (Ike 2), forming the IZP with an *oblique angle* collimated wave rather than a paraxial one. This has the effect of sustaining a more uniform image distance and consequential lower spot aberration over a wider range of diffraction angles from the hologram. Since the off-center zone lens segments are oriented along the periphery of the scanning disk, the illuminating beam is incident upon varying IZP radii during scan, causing a variation in diffraction angle Θ_o. Thus, monitoring of focal distance and quality as a function of Θ_o provides data as a function of corresponding disk rotation angle Φ. The output angle of the marginal rays at $(r \pm \Delta r)$ having grating frequency $\nu(r \pm \Delta r)$ may be derived directly from the grating equation [Eq. (4) of Section 3.2] as

$$\Theta_o(r \pm \Delta r) = \sin^{-1}\left[\nu(r \pm \Delta r)\lambda\right] - \tan^{-1}\left(\pm \frac{\Delta r}{a}\right) \qquad (48)$$

where the grating frequency and distance a are defined earlier.

Although no information of the relationship between Θ_o and disk rotation angle Φ is provided in their work, analytic and experimental data show reasonable spot uniformity over a Θ_o range of from 10 to 30° under the condition of magnification $M = 4$ from an IZP exposed with an oblique angle of 15°. Nonuniform velocity is observed, however, indicating approximately 20% change over the range 15 to 25° and almost a 40% change over a 10 to 30° range of Θ_o. In the intended POS application, this variation can be accommodated by electronic logic flexibility.

A more recent publication (Ike 3) indicates adaptation of the authors' oblique-angle exposure techniques to a different purpose—correction of the fringe shrinkage in silver halide (which occurs after wet processing) to restore higher diffraction efficiency. With the exposure angles selected—collimated $\Theta_i = 0$ and principal ray of spherical wave $\Theta_o = 50°$—a variation in efficiency is observed upon reconstruction with the same collimated wave while varying Θ_i. The maximum efficiency point is shifted by approximately 3°. By imparting a corresponding shift in Θ_i during exposure, the maximum is restored after processing when $\Theta_i = 0$, allowing reillumination with a paraxial collimated beam. A consequence of this procedure is to impart astigmatism to the reconstructed beam; of such values, however, that they are tolerated in the intended POS application. Interestingly, the authors in this more recent implementation have elected to operate at effectively unity magnification. A large scan arc is achieved more conventionally; by placing the axis of the holographic segment well outside the disk radius, as illustrated schematically in Fig. 2.2c.

It is noteworthy that the magnification $M = b(o)/f$ adapted by these and subsequent authors is unrelated to $m = \Theta/\Phi$ of Section 2.8.1. The first M is more like classic image magnification, independent of scan, while the second m relates to the ratio of output scan angle to the deflector rotation angle.

4.2.2.6.1.1. IBM POS CONFIGURATION. One of the most prominent convergent-beam disk configurations is that developed for IBM's POS supermarket scanners, described by Dickson and Sincerbox (Dic 1, Sin). It consists of an array of 21 holographic lenses oriented about the periphery of a 195-mm-diameter glass disk. Useful beam orientations and focal point variations serve effectively to illuminate a three-dimensional space with the scanning focal points, one of which is used for internal diagnostics. Additional note of this multidimensional scan is provided in Section 4.4.3.

The major departure holographically is the selection of $\Theta_i = 22°$ and $\Theta_o \approx 45°$; a compromise between the paraxial ($\Theta_i = 0$) illumination of the classic IZP and the $\Theta_i \approx 45°$ utilized for full Bragg operation (with $\Theta_o \approx 45°$) per Section 4.2.2.4. Since this is a convergent beam scanner, its spot aberration is a compromise between the perfection of $\Theta_i = 0$ illumination and the rather severe off-axis aberation (mainly astigmatism) when $\Theta_i \approx 45°$. Also, because of its noncritical POS application, fractional wobble sensitivity, per Eq. (28) of Section 4.2.2.4 is compromised at 0.31, compared to 0.41 at $\Theta_i = 0$ and its nulling at $\Theta_i = \Theta_o$.

Another factor that is determined by the selection of Θ_i and Θ_o is the scan magnification, m. From Eq. (29) of Section 2.8.4, when $\Theta_i = 22°$ and $\Theta_o = 45°$, then $m = 1.08$ in the central portion of scan. It is somewhat nonlinear and the scan is arcuate, but as expressed for the earlier work, these factors are well tolerated for POS application by electronic timing accommodation. Even the increased wobble sensitivity is uncritical because of the almost independent characteristic of each of the POS bar code scans. The important factors of resolution, speed, multiple scan, optical efficiency, and uniformity of return signal are implemented with careful control and allocation of the holographic facets, mirrors and their retro-collection areas.

4.2.2.6.2. Modified Zone Plates for Convergent Beam: Arcuate Scan

As innovative study by Ono and Nishida (Ono 2) approached the problem of disk scanning of a converging beam and minimizing its aberrations by formulating and solving an interesting generalization to the structure of zone plates (see Section 3.4.3). The authors start by analyzing the infinity-conjugate IZP [interferometric zone plate; see Section 3.4.1], developed with a spherical wave diverging from a focal point f interfering with a paraxial plane wave on a normal recording plane. Upon illuminating this holographic lens with a radially shifted paraxial but slightly diverging wave derived from a source distance greater than the original focal length f, a conjugate image point is formed by the diffracted converging beam. Its principal ray intersects the axis at f, and the surrounding marginal rays continue to form a conjugate image point some distance beyond f. Thus imaging properties similar to those from a finite conjugate lens or hologram are developed with one formed from an infinity-conjugate IZP. The basis for modification of zone lens fringe spacing arises from an analytic determination of this image point distance and its variation with scan angle.

Defining the same magnification factor M as the ratio of undeflected image distance $b(o)$ to the focal length f, the authors plot the variation in (normalized) image distance versus the diffraction angle Θ, with magnification as a parameter. In this finite-conjugate operation of an *infinity-conjugate* IZP, the principal input ray remains at $\Theta_i = 0$ while the principal output ray diffracts at Θ_o. Thus, when $M = 1$, the image distance remains constant, for it *is* the focal length, and a change in Θ_o reflects the change in radial distance r of the incident principal ray as $r = f \tan \Theta_o$. With increased magnification, however, the image distance increases rapidly with Θ_o, explained by the decrease in focal power of an IZP at increased radius (Lee 1).

The consequences of similar scan of an LZP (limited zone plate) are then considered. Identified by the authors as a GZP (geometric zone plate), the LZP notation is here preferred, as discussed in Section 3.4.1. The LZP exhibits fringe spacings determined from a small-angle approximation on the IZP. Without this "correction term," it hastens the compression of peripheral fringes and reduces the image distance with increased Θ_o, especially at increased magnification. This raises the interesting prospect of forming a generalized zone plate having fringe spacing adjustable between the limits of an IZP and an LZP. A simple variation on Eq. (1) for the zone radii appearing in Section 3.4.1 yields an expression with the new parameter N,

$$r_N^2 = 2nf\lambda + (n\lambda/N)^2 \tag{49}$$

where n represents the fringe number. When $N = 1$, this describes an IZP; when $N = \infty$, the second term drops out and it represents an LZP. Repeating the evaluation conducted above, now with a variation in N, it is found that at $N = 3$ and magnificiation $M = 4$, the image distance remains remarkably constant with diffraction angle Θ variation to $30°$.

Formation of these modified zone lenses is a challenge. Although they could be developed by computation and appropriate photoreduction of the plotted zones, the authors elected formation of the holograms by means of a novel but sometimes complicated interferometric technique requiring a sequence of holographic exposures. One can appreciate the significance of interferometric formation when realizing that a symmetric finite-conjugate zone lens is, in fact, one having $N = 2$, as represented by Eq. (10) of Section 3.4.2. Analysis by the authors, as depicted in their Fig. 4, shows that selection of this simple expedient provides reasonably uniform image distance with Θ_o variation to $30°$, at a magnification of 3.75.

N denotes the number of spherical waves applied to the hologram. The higher values of N are developed by judicious selection of successive stages of reillumination with different (spherical or collimated) wavefronts and reholographing their diffracted reconstructions (see Section 3.4.3). In this way, holograms of $N = 1$, 2, and 3 were made and tested. The $N = 3$ hologram exhibited conjugate foci of 200 mm and a focal length $f = 66.7$ mm and was tested with $M = 9$, forming the image distance of 600 mm. Spot size was 200 μm in the center of the $\pm 22.6°$ scan over an image length of ± 25 cm. Although the degree of spot enlargement over scan was not reported explicitly, it appeared not to exceed 20% of center spot size.

Complementing the notably good performance described above, its intended application accommodates some of its limitations. Targeted for POS (point of sale) scanning, scan bow and nonlinearity are not significant limitations—two factors prominent in this design, and thus discussed. Not indicated, however, is the wobble sensitivity, which is surely not minimized, for operation with the incident beam normal to the hologram is uniquely disparate from the Bragg condition (Section 4.2.2) and preference for a large diffraction angle to reduce bow accentuates this disparity. [See the subsequent discussion regarding Eq. (50).] It is important to reemphasize that these limitations are not serious for the intended application of POS scanning—which can tolerate significant bow, nonlinearity, and line misplacement.

The functional form of the prototype scanner is that of a seven-faceted disk, each facet providing successively displaced (arcuate) scan lines so that each rotation forms a seven-line raster. The holograms are $N = 2$, operating at $M = 4$. Note that $N = 2$ corresponds to a symmetric finite-conjugate hologram, hence a single holographic exposure per facet. The holographic segments are oriented about the disk such that their axes are parallel to the shaft but significantly beyond the disk; hence the diffracted output beams exit at a steep angle Θ_o with respect to the disk from the marginal segments of the zone lenses, where the fringe spacing is narrow.

4.2.2.6.3. Modified Zone Plates for Convergent Beam: Straightened Scan

Following their earlier work described in Section 4.2.2.6.2, Ono and Nishida (Ono 3) expanded their generalized zone lens approach to un-bow the scan line. This is achieved by two basic extensions: First, as indicated in the discussion above, the bow radius increases from operation with large disk radii and large diffraction angles. Second, increase of the parameter N—number of effective spherical waves forming the generalized zone plate (see Section 3.4.3)—for it was found that this tends toward a complementary bow. The greatest effect occurs in going from $N = 1$ to $N = 2$ (i.e., from infinite-conjugate to symmetric finite-conjugate zone lens construction). The effect tends to saturate by $N = 4$ (second order $N = 2$), exhibiting little increase at $N = \infty$ (LZP). Compromise occurs in a balance between the reduced bow of the larger scan radius and the complementary bow from the zone lens formed with higher N.

Under the conditions investigated, there was no need to form a holographic lens having N higher than 4. Experiment was conducted with $N = 4$ holograms, formed with multiple exposures from the second order of a "conventional" $N = 2$ hologram, as reviewed briefly in Section 3.4.3. Each final exposure (at 633 nm) was conducted directly on the silver halide master disk, which was then bleached and replicated.

Performance parameters (at a 633-nm reconstruction wavelength) were as follows: disk radius = 75 mm (at illumination center), focal length = 50 mm (made originally with conjugate foci at 200 mm), image distance (from disk and normal to disk) = 800 mm, $M = 800/50 = 16$, scan length = ± 200 mm with deviation from line straightness within ± 100 μm. Scan angle = $\pm 14°$. Spot size at center of scan ≈ 250 μm and enlarged somewhat astigmatically in the diffraction direction. The image surface was a plane normal to the rotating axis (parallel to the disk). Since the scan line is straightened, this limitation could be removed and the image surface placed in the most favorable angular orientation with respect to the beam axis. No attempt at such optimization was discussed.

As in the earlier work, this good performance was reported with no comment regarding wobble sensitivity. This factor is somewhat more critical with increased diffraction angle Θ_o, as seen from Eq. (28) of Section 4.2.2.4, when $\Theta_i = 0$ for the transmission hologram,

$$d\Theta_o \approx (1 - \sec \Theta_o) \, d\alpha \qquad (50)$$

in which $d\Theta_o$ is the change in output angle for a disk wobble error $d\alpha$. Also, extra care is required to provide uniformity of formation and dynamic centration of the lenticular hologram segments which form the facets. Once the operational objective moves from the earlier relatively relaxed repeatability requirement of POS scanners to the rigorous periodicity demands of line-scanned graphics, special attention must be exercised to constrain the effects of line misplacement.

4.2.2.6.4. Ray-Optimized Convergent Scan Near Bragg Regime: Straightened Scan

An exemplary analytic and experimental work was conducted (Iwa 2) in which the authors (Iwaoka and Shiozawa) investigated the optimization of a disk scanner which provides a convergent beam output. The objective was to create a system that requires no subsequent focusing, while meeting fairly rigorous criteria of spot image uniformity and scan straightness.* An additional object was

---

*The authors identify this characteristic as scan "linearity," a descriptor that normally represents geometric uniformity of pixel periodicity in the along-scan direction (constant scan velocity), following our discussion in Section 5.2.

to create the hologram simply with a He-Ne laser at 633 nm using nominal (nondistorted) spherical wavefronts and to reconstruct with a laser diode at 780 nm. Although these objectives were achieved to a high degree, and experimental performance corresponded closely to the analytic predictions, the system was left with the need for significant control over disk centration error. Also, because the optimization routine was limited to three factors of spot quality, scan straightness, and optical efficiency, no data are provided regarding sensitivity to disk wobble and degree of scan nonlinearity (see Section 5.2). One can expect, however, that optimization toward high diffraction efficiency would tend toward the Bragg condition, which would (by Section 4.2.2.4) also tend toward nulling the gross effect of disk wobble. Also, one can expect that scan linearity will be typified by the development in Section 5.2.

The scanning method is fundamentally similar to that described in Section 4.2.2.5 for straight-line scan from a plane grating. With the use of converging and diverging wavefronts rather that collimated ones, the gratings act, however, more like marginal sections of finite-conjugate Fresnel zone lenses (Section 3.4.2) than linear periodic structures. But the principal rays follow paths similar to those detailed in Section 4.2.2.5.

A carefully configured form of ray tracing was employed, in which a group of incident (reconstruction) rays distributed over their (diverging or converging) angles generates a one-for-one distribution of diffracted output rays which are to form the converging bundles intended to focus to the scanned point. Merit functions were established for each of the characteristics of aberration, straightness, and diffraction efficiency, limits were placed on tolerable deviations, and an iterative computational technique was employed to maximize a composite merit over the desired operating range. Since the scanner was to operate from a diode laser source with holograms formed on bleached silver halide with a HeNe laser, the wavelength shift from 633 nm to the nominal 780 nm was contained in the program (see Section 5.4). A complication of mode hopping of the diode laser was also investigated and reasonably resolved (discussed later).

Our review of the characteristics of this design is aided by reference to Fig. 4.30A showing principal rays of the input (reconstruction) beam incident on H at angle Θ_i and the output (diffracted) principal ray existing at angle Θ_o. The full output beam bundle (derived from a finite illuminated subtense on H) converges to focus on an image plane (not shown) that is mounted normal to the principal ray (when at center of scan). A major option expressed by the authors was reconstructing with a converging or a diverging beam. On our reference figure, a converging input beam would focus at a point some distance beyond H, and a diverging one would be derived from a point some distance before H.

That choice, as expressed by the authors, relates to the relative ease of forming one or the other from a diode laser source. Interestingly, no option was expressed for a collimated reconstruction beam, essentially midway between a converging and a diverging one (discussed subsequently). The diverging one is easiest to form, followed by the collimated and converging ones. The last one requires the largest final lens. Anamorphic beam normalizing (if to be used) can be similar for all. The diverging beam may be derived efficiently from a diode laser followed by a conjugate-imaging lens providing a new focal point having selected angular demagnification of the broad diode output distribution. The other two require a second lens to capture the first diverging beam and either collimate it or reconverge it. There appears, however, not to be commanding distinctions, as long as the converging or diverging f numbers are not too low. Hence judgment of choice can be modified from that selected by the authors.

The characteristics of the scanner, directed toward "laser printing," were as follows: scan format width = 300 mm; spot size (FWHM) $\leqslant$ 100 μm; output f number $\approx$ 100, active disk radius (to principal ray) = 40 mm. Distinctions between diverging and converging illumination were as follows:

1. Centration error tolerance was significantly more relaxed for the converging than for the diverging conditions. Numerically, the former was $\pm$92 μm, and for the latter, $\pm$2.8 μm (for transverse spot displacement of $\pm$20 μm); a 33-fold factor! This may be explained by the more uniform rate of change of ray angle at the hologram in the converging case (over the entire beam bundle), forming a grating having more uniform spacing. A perfectly uniform linear grating, for example, when illuminated with a converging beam, will diffract a correspondingly converging output beam, albeit aberrated. The problem is, therefore, to select the exposure orientations that provides tolerable aberration.

2. Spot aberration was lower in the diverging beam case than in the converging one; although even there, spot sizes approached the 100-μm FWHM objective.

3. Line straightness was very similar in both cases ($\approx$ 100 μm deviation), with a slight (10%) edge in favor of the converging beam illumination. Interestingly, and not reported explicitly, the converging beam case also provides somewhat lower ($\approx$ 10%) angular scan magnification than in the diverging beam case (i.e., larger rotational angle for the same scan angle). Note that the converging beam case approaches more the condition of radial symmetry, hence closer to unity magnification.

Except for centration error, the distinctions are not too profound. The authors preferred the lower spot aberration and greater ease of implementing the diverging beam configuration. However, it requires intensive mechanical control of the rotating member to constrain spot misplacement due to eccentricity.

It is instructive to consider the effect of exposing and reconstructing with a collimated beam. A first-order expectation is that performance will be somewhere midway between the two conditions analyzed. This sounds like a good compromise, especially with the significantly relaxed centration tolerance that is sure to result. In fact, a good approximation is that centration error becomes somewhat better than linear—that is, perhaps 25 μm eccentricity for 20-μm spot misplacement. With a collimated beam, small translation of the beam will not displace the diffracted spot. Conversely, small transverse displacements of the disk will carry the spot along with the disk. If the image plane were at 45° to the axis, the displacement would be reduced by $\sqrt{2}$; or 28.3 μm eccentricity for 20 μm spot displacement—a 10-fold relaxation over the 2.8-μm sensitivity of the divergent beam case.

During initial test of the experimental system and modulating the laser diode, its mode hopping (wavelength shift due to change in average current—hence temperature) caused excessive spot misplacement in the cross-scan direction. The magnitude of this displacement is expressed as

$$\Delta y = \frac{f\nu}{\cos \Theta_o} \Delta\lambda \tag{51}$$

where f is the distance from hologram to image, ν the spatial frequency of the hologram (in the y direction), Θ_o the diffraction angle, and $\Delta\lambda$ the wavelength shift. For a 0.5-nm shift in this system, this calculates to a 0.2-mm spot misplacement. An effective solution was to radio-frequency modulate the laser diode about a midway direct-current bias, allowing the power differential between modulated and unmodulated to approach zero. Section 5.4.4 provides additional consideration of this wavelength-shift problem and another method of correcting this effect.

4.3. TRANSLATIONAL SCAN

4.3.1. Introduction

Although most of the earlier discussion related to the rotation of a diffractive member about an axis, it is clear that one may also translate a (lenticular) diffractor and appreciate a corresponding movement of the image point. Some of the characteristics of such translational scan were introduced in Section 2.8 during our discussion of resolution. It was there expressed that a focused spot of size δ which is conveyed over a continuous distance S accumulates resolution as $N_s = S/\delta$; directly, the number of spots disposed (with separation $w = \delta$)

within the scanned distance S. Since S may be a curvilinear path and w_N may
be a set of N nonuniform displacements such that $w_N \neq \delta_N \neq \delta$, the resolution
is regarded simply as the number N of spots along that path. Such is the case
for some discontinuous and curvilinear scan systems appearing in this section.
Also, when $w \gg \delta$, spot aberration is often considered insignificant, while
when $w \approx \delta$, spot aberration (and misplacement) can be crucial.

A basic requirement for scan of an image point in one direction is that the
translated component need exhibit angular power in that direction. For example, a (positive) cylindrical or holographic lenticule, when oriented with power
in the x direction (providing no power in the y direction) and illuminated with
z directed collimated light will form a focused line along y which will translate
along x when the lenticule is so translated. If translated along y, no effect will
be observed as long as there is no vignetting. If the cylindrical lens is rotated
about z through a fixed angle β of up to (but not equal to) 90° and then translated
along x, a corresponding translation of the diagonal (at angle β) focused line
will occur along x. Only the resolution will be reduced, for the x-resolved line
width increases as $\delta_x = \delta_o \sec \beta$, where δ_o is the line width in the direction of
lenticular power. Correspondingly, a plane linear diffraction grating (PLDG)
provides no changing optical power, only a fixed angular shift to the incident
illumination. Thus, if translated within a fixed illuminating field, it exhibits no
spatial scan (see Section 3.2), independent of the manner of illumination. Utilization of the PLDG for scanning requires angular orientation or changing the
period of the grating structure.

How, then, is translational scan to be utilized without mounting the component on a rotating member? Functionally, with little advantage over the translation of equivalent conventional optics. As is well recognized, classical optical
elements are seldom translated for scanning applications. Although examples
are found in reprographics and in other relatively slow-moving rack/track-con-
strained instruments, the problem is the running-out of linear translation space
and requiring reciprocation or tape and belt drive. The concept of lenticular
translation is useful for expressing fundamental properties and for describing
important variations, some described further in Sections 4.4 and 4.5.

4.3.2. Large-Aperture Translational Systems

4.3.2.1. Early Work

One such variation is expressed in the independent and joint work by Gerbig
(Ger 1, Ger 2, Ger 5) and Case (Cas 2, Cas 3). The purpose was to provide
preprogrammed two- or three-dimensional (x, y, and z) positioning of a point
along an arbitrary path employing only linear translation. The holographic ele-

ment is anamorphic, as is the cylindrical example above. The problem is to generate the grating structure (by appropriate computation and fabrication) such that the grating periodicity (forming anamorphic power and "wedge") and its angular orientation change properly along a strip of substrate material. Figure 4.36 illustrates (Ger 1, Cas 2) the principle. The hologram is provided with a structure that approximates, over small displacements, linear gratings oriented at varying angles and periodicities. The first-order consequence of its translation in x (in x-y plane) is an angular change in the diffracted output beam which, when focused by the lens, forms focal spot movement along an arbitrary path in the v-μ plane.

The prospect of such operation was expressed earlier (Bry 1), including encoding the gratings with lenticular changes (forming beam convergence or divergence) corresponding to focal shifts—an extra dimension to positioning the scanned focal point. A simple linear scan is expressed in Fig. 4.37, a reproduction from that work by Bryngdahl. Figure 4.37a shows translation of a central strip of a positive zone lens element which forms image point movement along the axis, analogous to lenticular operation in Fig. 2.6. Figure 4.37b shows translation of an off-axis strip from the same type of diffractor so that the image point moves parallel to the axis but is displaced vertically to separate it from the (axial) zero-order component. Finally, in Fig. 4.37c is shown the transformation of the grating function (Fig. 4.37b) onto the periphery of a disk to simplify repetitive scan by rotation about an axis. Subsequent publication by Bryngdahl and Lee (Bry 2) concentrated on holograms on rotational substrates, reserved for coverage in later sections (4.4.3.1 and 4.5) on computer-generated diffractive scanning. An interesting adaptation of zone lens translation is the generation of a raster with a contiguous array of zone lens strips (Ger 1).

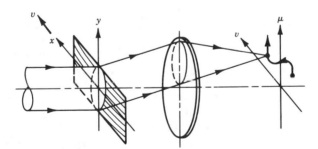

Figure 4.36. General translational holographic scanner. Hologram in x-y plane is shifted along x at velocity v, changing angle of output beam focused by lens to arbitrary path in v-μ plane. (From Ref. Ger 1.)

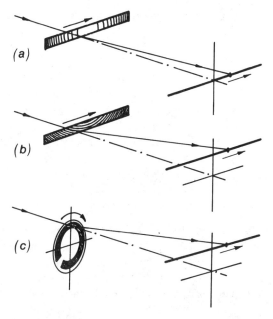

Figure 4.37. Computer-generated hologram line scan arrangements. (*a*) and (*b*) Zone lens sections in translation. Off-axis strip of (*b*) is transformed to rotation in (*c*). (From Ref. Bry 1.) North Holland Physics Publishing Co., 1975.

4.3.2.2. Attendant Aberration during Arbitrary Scan

For the type of holographic strip which is to impart a general (arbitrary) displacement of the imaged focal point, it was appreciated (Ger 1) that if illuminated over a finite aperture in the translation direction, one cannot create continuous scan without attendant aberration (Cam). The problem can be visualized effectively by considering the strip formed with two (or more) adjacent gratings, each providing high-integrity diffraction into adjacent image points. The gratings differ only slightly from their neighbors. When the beam illuminates one grating, one stigmatic image point results. When the strip is translated to the next grating, the adjacent image point appears properly. During translation, however, when the beam illuminates portions of both gratings, both images are represented simultaneously by the truncated apodization of each grating. A continuum of such grating structure on a strip, when illuminated conventionally and translated, can be made to diffract only the centroid of the image points along the selected arbitrary path. Image point aberration will depend on the spatial rate of change of the grating structure with respect to a given illumination subtense—the slower the change, the lower the aberration. The effect is analogous to that of a reflective strip of tape which is continuously distorted such that the average reflection of a laser beam is positioned as desired. The slower the change for a given aperture size, the longer the tape and the more rapidly it must be translated for a given scan time. The aberration develops for scan of an arbitrary pattern. For scan of a regular pattern, such as a straight line, a segment of a Fresnel zone hologram will serve well, as represented in Fig. 4.37 and discussed in Section 4.5. Or a translating one-dimensional zone lens can be provided with quadrature focusing with a tandem fixed cylindrical lens having positive power in the y direction. Displacement of proper zone or holographic lenses (see Section 3.4.1) provides stigmatic scan.

4.3.2.3. *Translational Focus Variation*

Variable optical power can also be provided by translational components; the result of elegent analytic and experimental work, as described by Lohmann (Loh 1). Refractive, Fresnel zone, or kinoform (Jor) lenses are so configured that when a pair is overlapped and shifted (in opposite directions by equal amounts), the pair operates as a refractive or diffractive cylindrical lens of continuously variable optical power. The diffractive elements (Fresnel) or kinoform) are formed by moiré interference of their binary structures. Two such pairs adjusted to equal power and mounted in tandem and in quadrature form the equivalent of a rotationally symmetric lens. Lenticular power may be varied by symmetric translation of the four elements while maintaining paired quadrature orientation. While similar variable moiré techniques might be considered to form variable periodicity gratings for scanning, the linear translation of single preformed diffractive elements is operationally more efficient, as interpreted by several investigators discussed here and in Section 4.4.3.

4.3.2.4. *Translational Scan of Interferometric Holograms*

As manifest in more recent work (Cas 2, Ger 5), a contiguous array of individual interferometrically generated thick holograms forming PLDGs (plane linear diffraction gratings) having differing grating spacings and varying angles to the x axis was considered as a useful variation to Fig. 4.36. It was tested as a discrete spot position scanner (Cas 2, Cas 3) (as though illuminated by a pulsed laser) to access any point on a 5×7 matrix for alphanumeric character generation. This takes advantage of the high diffraction efficiency of the thick (dichromated gelatin) transmission hologram. While the image points in the 5×7 array were separated by about six times the spot diameter, good individual spot integrity was provided.

  In another experiment aimed at forming a contiguous set of dots (Cas 2, Cas 3), a preprogrammed array of 64 such holograms was arranged on a strip, individually illuminated, and indexed in sequence to form a symbol (a question mark) whose dots were ideally separated by their Rayleigh distance. Although the character appeared quite presentable, this preprogrammed method relinquished the privilege of random access.

To provide almost random access to a higher-resolution dot matrix (16 × 16 = 256 dots), a tandem arrangement (Ger 1) of two 16-position transmission holograms was formed (Cas 2, Cas 3) and demonstrated a very respectable regular array of 16 × 16 addressable dots. The tandem arrangement allows (pulsed) selection of any one of the product of elements provided by each. Another implementation of two tandem holograms (Cas 2, Cas 3) utilized the first as the 5 × 7 x-y point scanner described above, and the second as a set of holographic lenses, each of varying power to provide adjustable depth to the focal point—thus dot positioning in x, y, and z.

4.3.3. Narrow-Aperture Translational Scan

To scan a contiguous arbitrary pattern having near-stigmatic spot quality from continuous hologram motion, anamorphic optics was added (Ger 4, Ger 5) to reduce the perturbation of data in the translation direction to near zero. The basic method is illustrated in Fig. 4.38 and may be compacted with replacement of transmissive optics with reflective components and the addition of optical power in the hologram. By limiting (anamorphically) the illuminating subtense horizontally to a narrow (vertical) line, the information extracted from the hologram is uniquely vertically oriented; there is no horizontal component. Note the use of anamorphic optics in a manner similar to that for reducing the effect of multifacet deflection errors in the cross-scan direction (described subsequently in Section 5.1). Thus with translation of the hologram horizontally is imparted spot deflection vertically in the scan plane. The holographic "tape" is now much shorter and the encoding requirement for generation of the diffraction pattern is more efficient. That is, for computer generation of an arbitrary two-dimensional set of $N \times N = N^2$ spot positions, a minimum of N^2

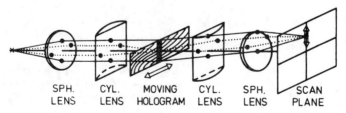

Figure 4.38. Translational holographic scanner utilizing anamorphic optics to reduce perturbation in one direction. (From Ref. Ger 4.)

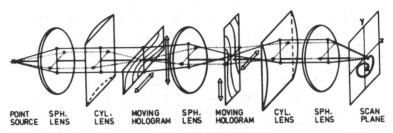

Figure 4.39. Tandem arrangement of two translational holographic scanners coupled through anamorphic optics to provide two-dimensional spot positioning with low aberration. (From Ref. Ger 4.)

cells is normally required in each hologram to conform to space–bandwidth conditions (Lee 2). For this one-dimensional scan, a minimum of N (rather than N^2) cells is required along the narrow vertical stripe on the holographic medium, each stripe arranged to diffract the beam into the desired vertical position.

To provide two-dimensional scan with low aberration, Fig. 4.39 illustrates (Ger 4, Ger 5) a tandem arrangement: one for vertical spot motion as before, and a second for horizontal spot motion. With appropriate hologram and translational programming, such a system will provide a general $(x\text{-}y)$ scan pattern of an imaged spot, with aberration determined primarily by the quality of the components rather than from the fundamental constraint of the illuminating aperture overlapping elements having conflicting diffractive properties. Section 4.5, on computer-generated diffractive scanning, discusses several related characteristics and processes.

4.4. MULTIPLEXING AND MULTIDIMENSIONAL SCAN

The purpose of multiplexing is to derive from a single facet, or from a group of related facets, output beams that are directed into different paths, and/or are focused over different lengths, and/or diffracted at different wavelengths. Multiplexing may be instituted in two principal forms, to provide

1. Different output beams from the same facet,
2. Different output beams from multiple facets.

In the first case, a single facet is illuminated with multiple-characteristic beams such that the diffracted output beams are differently represented, either in position or in wavelength. In the second case, each of a group of adjacent facets is illuminated with individual beams such that each facet redirects its output beam with desired differing characteristics. Although multiplexing by the first case is functionally possible, there are few examples of such operation with holographic scanners. The second case, exemplified by Fig. 4.17, is more prevalent.

4.4.1. Multiplexing from the Same Facet

In holographic scanning, some limitations constrain multiplexing by method 1. Consider, first, the hologram requiring a unique angle of illumination, as represented by Fig. 4.9. Here a disk substrate requires radially symmetric illumination in the form of a diverging wave derived from point p_o' on the rotating axis. This configuration requires a holographic facet exhibiting optical power, as do many others, exemplified by Figs. 4.22, 4.23, 4.27, and 4.28. In spatial multiplexing the output beams are directed to different locations. Thus the different input beams need derive from points adjacent to p_o' of Fig. 4.9, all on the axis to sustain radial symmetry. A multifrequency-driven acoustooptic modulator will, for example, provide a fan of closely spaced beams that approach this condition. Due to the narrow field angle of some holographic elements, aberration of all spots reconstructed from points other than p_o' need be considered. Also, to focus at a different location (Ish 1), the illuminating point is withdrawn from p_o' to a point off the rotating axis, modifying radial symmetry of the output scan function. One may consider the use of nonlenticular (plane linear) gratings with subsequent focusing to reduce these constraints in a particular scanner configuration.

The spatial multiplexing described above is further complicated by wavelength multiplexing—seeking to impart to a single-output (scanned) focal point a superimposition of radiations of differing wavelengths, such as for scanning or recording of (primary) color information. If the hologram medium is sufficiently thick (grating storage throughout substantive volume), it may sustain wavelength selectivity during reconstruction (i.e., suppress all but the desired ones). Under this condition it is possible to multiplex a small group of beams of differing wavelengths (say, three primary colors) with superimposed stigmatic reconstruction at the same focal point, reilluminated from the same point, such as p_o' of Fig. 4.9. Although adaptable to stationary holograms, to our knowledge, this has not been implemented in scanning. A proposal related to this approach is by Locke (Loc), where, represented by Fig. 4.9, his patent proposed illuminating a "thin" reflective relief hologram with such multiple-wavelength beams and depending on spatial filtering at the output to separate unwanted diffracted components. A thin hologram will diffract each of the three incident beams into (at least) three different locations, one desired and the others undesired. Thus the outputs from three different wavelength inputs will consist of three superimposed desired beams and many undesired ones which are separated spatially from the desired cluster. The method of spatial filtering in the image plane for extraction of the desired scanned image point is not described by Locke. Perhaps a "slit" window in the image plane can provide requisite filtering of the unwanted components—to be determined.

Another complication to multiplexing spectrally in the same hologram is the reduction of radiometric efficiency. Sequential or multiwavelength exposures superpose incoherently. Even if the fringes are distributed within an ideal volume hologram, each reconstructed wave suffers a loss of radiometric efficiency approximated by $1/N_s^2$, where N_s is the number of superposed holograms (Smi). If exposed simultaneously coherently, the efficiency of each is reduced by the approximate factor $1/N_s$. If the hologram medium is thin, as exemplified above, additional radiometric loss is manifest in the unwanted diffracted orders, which require obscuration by spatial filtering.

If multiplexing from the same facet is operationally important, this is one of the principal disadvantages of scanning holographically. No serious spectral or efficiency limitations exist in pure reflection from plane facets. In holographic scanning, those problems that arise due to lenticular facet angular and wavelength selectivity can be alleviated in part by the use of plane linear gratings. But if angularly separated, the multiple beams may not execute adequately congruent (or parallel) scans. And if fringes are superimposed, we are left with a sacrifice in radiometric efficiency, at least of magnitude $1/N_s$ (if coherent), due to fringe amplitude sharing of $1/N_s^{1/2}$ each of the individual exposures.

4.4.2. Multiplexing from Different Facets

Almost all of the problems discussed above disappear when each facet is dedicated to a single task. Each holographic exposure can be optimized for its most effective reconstruction. A residual problem exists in nonradially symmetric systems (such as the linearized disk scanner operating near the Bragg regime, by Section 4.2.2.5), where the scan linearity is a critical function of the selection of input and output angles (Sin). If radially adjacent facets are used, they may be sufficiently displaced to alter these angles sufficiently to perturb the congruence or integrity of the multiplexed set of beams.

A form of multiplexing which can be moderately sustained is that of focal-length change from angularly adjacent facets in focusing-type systems, by Section 4.2.2.6.1.1. In this point-of-sale application, the scan linearity is noncritical, and any nonuniformity of the scan function due to angular change resulting from focal-length change is well tolerated.

There appears, however, no inhibition to multiplexing from different facets in the radially symmetric systems, as represented in Fig. 4.17. Each reconstruction can be optimized while scan integrity is sustained. This is exemplified more recently in a proposal by Ishikawa and Noguchi (Ish 4) which derives a separate pilot beam for scan synchronization from a set of hologram facets adjacent to the main scanning ones. It is also represented in a simple form by Hecker (Hec), in which rotation of a set of segments of coaxial zone lenses on a disk shifts the diffracted focal points incrementally along the rotating axis. Another arrangement of two concentric rings of differing holographic facets on a disk (Ih 7) provides combined low- and high-resolution scans.

4.4.3. Multidimensional Scan

The most prevalent multidimensional scan is that in the x-y plane, usually in raster format. A third dimension might be added with focal-length accommodation, per prior discussion.

The first diffractive optics scanner providing two-dimensional area coverage was invented in 1931—even before the introduction of the concept of holography by Gabor in 1947. This remarkable work of Hollis Baird (Bai) is further represented and illustrated in Appendix 1. Although primitive, the concept is directly transferrable to holographic scanning, as by the patent filings of McMann (McM 4) in 1967 and by McMahon (McM 2) in 1968. Portions of both works are described further in Appendix 1. Interestingly, the McMann filing was frustrated by Baird, as was the portion of the McMahon filing relating to two-dimensional scan from a single rotating element, such as that patented by Baird. Only the innovation that relates to the use of a pair of rotating holographic disks in tandem, providing three-dimensional scan, survived in the three claims. Figure 4.40 is a reproduction of Fig. 4 of the McMahon patent. The scanned output from the first element 58 forms a variable input characteristic to the second one on disk 55—potentially compromising image quality due to changing illumination on the second hologram, per our discussion in Section 4.4.1. The consequence of this action merits attention even if relay techniques (Bei 6) are employed (per Fig. 4.40, where lens 53 images the illuminated aperture of hologram 58 upon the hologram of the second disk). If 58 is over-illuminated, the image of 58 will scan across disk 55. If grossly underilluminated, so that a small beam is incident on hologram 58, its image will—as desired—be immobilized by lens 53 on the second disk. But in both cases, the angle of incidence of the illumination on the holograms on disk 55 will change during scan of the first disk, exposing the final diffracted output to geometric error and aberration. The degree of tolerance of this action is, of course, a function of the intended use of the system.

When more than one raster is to be derived from the same scanning device, such as crossed rasters applied to bar code scanning, a technique adapted from conventional scanning may be used quite effectively. Figure 4.41 illustrates an implementation by Charles Cheng (Che 1). A single laser beam is split into two; each directed to a different facet on a holographic disk which generates a single raster per revolution per beam. Output beams 75 and 76 will be scanned

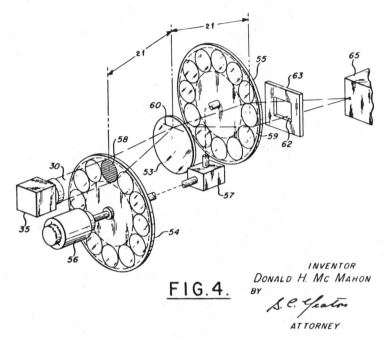

FIG. 4.

INVENTOR
DONALD H. MC MAHON
BY
ん.ℓ. やﾞe╱ん
ATTORNEY

Figure 4.40. Reproduction of Fig. 4 from U.S. Patent 3,619,033 by Donald H. McMahon entitled, "Three-Dimensional Light Beam Scanner Utilizing Tandemly Arranged Diffraction Gratings." Relay lens 53 is positioned a distance $4f$ from each scanning disk 54 and 55, which form three-dimensional scan in volume 65.

approximately orthogonally with respect to each other. Mirrors 92 and 94 redirect the two rasters to the same region upon the bar code surface 32. The backscatter radiation is rediffracted by the holographic facets, descanned thereby, and directed by mirror 80 through a collecting lens 96 upon detector 98 for signal extraction. This retrocollection is in a manner similar to that described in Section 4.2.1.3.2 and utilized by IBM in the POS disk scanner system discussed in Section 4.2.2.6.1.1. The apparent grating orientations of the holograms in Fig. 4.41 are illustrative only. They are (focusing) lenticular holograms with gratings as marginal sections of zone lenses oriented in approximate quadrature to those illustrated.

A variety of multiple-scan implementations from different facets are represented in a group of POS designs by Ikeda, Ando, and Inagaki (Ike 1), introduced in Section 4.2.2.6.1. This patent provides a veritable tutorial on the design of several interesting techniques. One provides a "stitch-bar" pattern (single long line crossed by an array of short lines); another, a "zigzag-bar" pattern (single long line crossing a sawtooth scan), and a crossed grating such

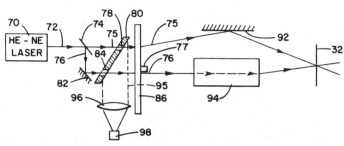

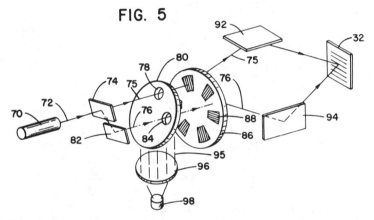

Figure 4.41. Reproduction of Figs. 4 and 5 from U.S. Patent 4,444,509 by Charles C. K. Cheng entitled "Holographic Scanning System." The output beams 75 and 76 form a pair of rasters in a crossed-grid pattern for bar code scanning on surface 34. Apertures 78 and 84 in mirror 80 allow passage of illuminating beams, while full mirror reflects the backscattered and rediffracted illumination toward detector 98.

as that described above. Some are split-beam illuminations upon different areas of the rotating holographic disk, as by Cheng above, while others utilize single beams. One single-beam approach utilizes a rotating disk with holograms of two general angular directions; one type forms the long "bar" scan, while the other directs its beam to a dove prism that rotates the resulting short scans by approximately 90° to form the "stitches" that cross the bar. Another uses a pair of overlapping counterrotating disks having alternating holograms and holes, phased such that a single beam either passes or is diffracted alternately, forming an image array of crossed x's. Still another single-beam approach, although potentially feasible, will not operate as specified and illustrated. Here the authors reverse the typical procedure by maintaining the holographic disk *fixed* and rotating the illuminating beam around the disk, effectively *translating* the beam across the zone lens holograms. However, the input beam in their Fig. 11 appears collimated, where such translation (with respect to the fixed hologram) will not effect movement of the image point, as desired. Instead, a finite conjugate point (in this case, a diverging input beam) must be translated to form a translating image point (see Section 2.4).

Another technique for deriving area and volume coverage from the same scanning device is that used effectively by IBM in their POS holographic scanner introduced in Section 4.2.2.6.1.1. Each facet is designed to direct the output scan to a different location—some to one mirror which leads to one scan pattern, and some to another mirror which provides other crossing scan patterns—the composite effectively filling the test volume with multidirectional and multifocused scanning beams. While such diversity of output scans can, in principle, be derived from a reflective rotating polygon having unique facet angles and lenticular powers, the economic virtue of conducting this task holographically (in high production) is quite clear, especially when the holograms also provide spectral filtering for the retro-directed signal.

A clear alternative to providing the additional dimensions of scan by holographic means, as described thus far, is by following a holographic scanner with a conventional one, such as a nutating and/or translating mirror. As in conventional scanning, care need be exercised in accessing the output of the first scanner and conveying it to subsequent ones; sometimes assisted with the use of relay optics (Bei 6) to avoid enlargement of subsequent apertures. If a subsequent scanner is holographic, the additional constraint of changing incidence angle must be evaluated, as expressed earlier in the discussion relating to Fig. 4.40. This constraint is effectively eliminated through the use of plane mirrors as subsequent scanners. Curved mirrors impose the extra consideration of changing lenticular orientation and potential resulting image degradation.

A good example of this option is represented in a work by Ih (Ih 5), in which is proposed the shifting or tilting of an auxiliary reflector such as that represented by AR in Fig. 4.28. This reflector now becomes a convenient second deflector for positioning the beam in a direction generally orthogonal to its original scan direction, for the purpose of providing field coverage, such as that of a raster. Some detailed effects of these actions merit attention.

Viewing the shifting option first, consider translating the AR of Fig. 4.28 (side views) parallel to the axis. Radial symmetry is maintained, while the output beam O is intercepted by the AR at a different average slope, causing (desired) angular change of the reflected beam (in the plane of the paper, side views). If the output beam is focusing, the focal point will also shift, to accommodate the change of length of beam O as intercepted variably by the AR. Now, viewing the tilting option, if one pivots the AR about a point that is the intersection of beam O with the AR, the focal path length remains essentially constant, while the deflected beam is changed angularly (through an angle $\approx 2x$ that of the AR changing angle), in the same plane as before. However, radial symmetry is compromised. In both cases the reciprocal action of the AR with respect to its original interposition during exposure of the holograms is also disturbed (see Section 4.2.2.3). The degree and severity of these ancillary effects is, of course, a function of the degree of change and the tolerance by the system to distortion and aberration (Sin), probably fully adaptable to such applications as POS scanning.

4.4.3.1. Scan in Direction Other Than Hologram Movement

A principal work by Bryngdahl investigated light deflection using computer-generated diffraction elements (Bry 1), revealing that to scan at a uniform rate from a uniformly articulated hologram, it is necessary for v_x, the grating spatial frequency in the x-scan direction, to change as a linear function of x. This is achieved when the phase function is of quadratic form,

$$\Phi(x, y) = \pi(x^2 + y^2)/f\lambda \tag{1}$$

recognized as the small-angle approximation for the circular zone lens represented by Eq. (7) of Section 3.4.1, in x-y coordinates.

Subsequent work by Bryngdahl and Lee (Bry 2) showed means for scanning in directions other than the direction of hologram movement in several novel computer-generated implementations. In one, a stitch-bar pattern is formed with unidirectional scan. In another, a crossed triangular pattern is developed; and in another, a spiral scan is generated with unidirectional movement of the hologram mounted on a drum support.

Generalizing further, Ono and Nishida (Ono 4) investigated different grating phase functions to yield the conditions under which image point movement can differ from hologram movement direction. They showed that some requisite phase relationships can, indeed, be generated interferometrically, circumventing the earlier expectation of requiring computation and plotting. In a form similar to Eq. (1), a more general phase distribution for circle, ellipse, and hyperbola is expressed by

$$\Theta(x, y) = \frac{2\pi}{\lambda}\left(\frac{X^2}{a^2} \pm \frac{Y^2}{b^2}\right) \tag{2}$$

The positive sign represents the circle ($a = b$) and the ellipse ($a \neq b$). The negative sign represents the hyperbola; point symmetry when ($a = b$) and coordinate symmetry when $a \neq b$. With $a^2 = 2f_x$ and $b^2 = 2f_y$, it reduces to Eq. (1). Instituting a coordinate transformation of rotation by Θ in which

$$X = x \cos \Theta + y \sin \Theta$$
$$Y = -x \sin \Theta + y \cos \Theta \tag{3}$$

the phase gradients on the y axis are determined and an angle φ is derived, representing the angle between the scan vector and the x axis. It is expressed in the form of tan φ, for which the numerator is shown to be $b^2 \sin^2\Theta \pm a^2 \cos^2\Theta$. When the numerator is zero, φ is zero. This occurs when the phase distribution is hyperbolic, when $b^2 \sin^2\Theta = a^2 \cos^2\Theta$. This equation reduces to the interesting condition, tan $\Theta = a/b$. That is, the tangent of the rotation angle is equal to the slope of the asymptote. For example, when $a = b$, $\Theta = 45°$, we form a phase distribution represented by Fig. 4.42. This is a reproduction of Fig. 7 from Ref Ono 4 showing a magnified central portion of a near-point-symmetric hyperbolic fringe pattern. It provides a clear representation of the asymptotes at $\approx 45°$. When the axes are oriented at this angle Θ [per Eq. (3)], translation of the grating along the new Y axis (with respect to a fixed illuminating beam normal to the plane of the grating) registers primarily as a grating spacing variation in X—that is, deflection in X ($\varphi = 0$) for grating translation in Y.

Following a presentation of means for generating the hyperbolic and elliptic phase functions, the authors (Ono 4) discuss a problem of anisotropy of the output beam diffracted from a noncircular phase grating—divergent in one direction and convergent in its quadrature direction. Novel means are proposed

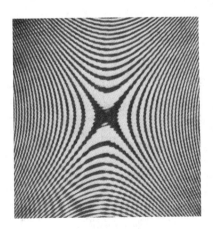

Figure 4.42. Magnified central portion of hyperbolic interference pattern with asymptotes at approximately $\pm45°$. (Reproduction of Fig. 7 from Ref. Ono 4.)

for complementing this anisotropy with a tandem second hologram situated on the opposite side of a drum scanner; the input beam is diffracted at the first scanning hologram and then corrected with the second (cylindrically) converging hologram. This interesting technique was demonstrated experimentally, to provide several forms of intersecting scan lines on an image surface, derived from drum scanners illuminated with a single beam and having no repositioning mirrors or prisms to reorient the scan direction.

For diffractive scanning in completely arbitrary directions by grating movement in one direction, see the work of Gerbig and Case discussed in Section 4.3.

4.5. COMPUTER-GENERATED DIFFRACTIVE SCANNING

4.5.1. Introduction

Since diffractive elements function by virtue of their near-periodic discontinuities called gratings, it is clear that if one forms the dominant spatial characteristics of these discontinuities by means other than interferometry, they, too, will serve as diffractive elements. Thus there is a rich background of use of mechanical means for forming gratings of regular structures, such as plane-parallel grids and plane concentric ones—the "plane grating" and the "zone lens," respectively. With the advent of rapid computational and precise plotting capabilities and a keener understanding of the nature and structure of these phase discontinuities, the use of computer-generated holograms (CGHs) applied to optical scanning was developed. A comprehensive work by W.-H. Lee (Lee 2) provides a noteworthy review of the development of the CGH, including its application to optical scanning. The history of the CGH since its introduction by Brown and Lohmann in 1966 (Bro) includes several ingenious variations in coding techniques for representing the amplitude and phase of the effective interfering wavefronts. These include the original detour phase hologram (Bro), modified off-axis reference beam holograms, kinoforms, and computer-generated interferograms (Lee 2). Interestingly, the most representative coding technique for scanning application is the last one—forming a high-density grating structure which emulates that of an interferogram which may not be physically realizable by wavefront interference using conventional wave-shaping procedures.

As introduced in Section 4.4.3.1, the early work in CGH scanning utilized primarily the equivalents of collimated and spherical wavefront interferences. When both are collimated, the plane linear grating is formed. As developed in Sections 3.4.1 and 3.4.2, when one is collimated and one spherical, this generates the infinity-conjugate zone lens, and when both wavefronts are spherical and in-line, the finite-conjugate zone lens is formed. The plane linear grating can scan only by its rotation with respect to an illuminating wavefront; it is translationally invariant. The latter two can scan by rotation (about a noncentered axis) or by translation. Clearly, rotation about their centered axes yields no scanning. Rotation about a noncentered (infinity-conjugate) zone lens axis is represented by Fig. 4.24 (and its transmission equivalent). Most scanned finite-conjugate holograms are off-axis ones, exemplified by Figs. 4.10, 4.22, 4.23, and 4.27. The reference/reconstruction wave is typically concentric with the rotating axis while the object/diffracted wave is directed from/to a point displaced from the axis.

In interferometric hologram scanning, we seldom investigate the detailed grating structure. The gratings are formed optimally when exposed properly. The principal interest is the range of grating periodicity, to be compatible with its formation in the storage medium, and for achievement of desired efficiency and its uniformity over the range of illumination conditions. The grating period is determined directly by application of the grating equation [Eq. (4) of Section 3.2] over the limiting ranges of marginal ray angles.

For computed grating formation it is essential that the grating structure be defined explicitly over its full operating region. Since there appears little incentive to scan computer-generated holograms of simple plane linear gratings or zone lenses (formed easily interferometrically), having well-defined phase-change grating equations [e.g., Eq. (1) of Section 4.4.3.1], grating functions must be determined which allow execution of more complex scan patterns. This is exemplified by the more general conic section phase distributions of Eq. (2) and its coordinate transformation of Eq. (2) of Section 4.4.3.1. Since these functions, too, may be generated holographically, as are the even more complex ones discussed in Section 3.4.3, this leaves for computer generation a special group that may not be formed interferometrically.

4.5.2. Piecewise Deposition of Interferometric Gratings

A combination of both disciplines is represented and discussed earlier in Section 4.3.2.4, in which a set of interferometrically generated plane linear gratings of differing grating spacing and angular orientations are deposited on a translatable strip such that each hologram diffracts into a different angle (Cas 2). When focused by an objective lens, the resulting pattern traces either a piecewise set of points forming a curvilinear path or a set of discrete spot positions which form alphanumeric characters. This technique, using thick holographic gratings, retains the advantages of high diffraction efficiency into the first-order and point positional accuracy determined by controlled object beam angular orientations during holographic exposure.

4.5.3. Conformal Transformations

One of the most important grating function transformations for optical scanning is that conducted in the earliest work, by Bryngdahl (Bry 1), introduced in Section 4.3.2.1 and illustrated in Fig. 4.37. To overcome the awkwardness of linear translation for providing continuous recurrent scan (the need for either long "tape" holograms or reciprocating motion), the zone lens grating for translational scan was transformed conformally to the marginal diameter of a disk. This allowed rotation of the disk to generate periodic linear scan of a focused point. Figure 4.43 illustrates the zone lens sectors (a) on-axis and (b) off-axis and the transformed annular band (c) corresponding to (b), which form the line scanners represented in Fig. 4.37. When annular band (c) is mounted on a disk and rotated about its center, it provides the same scan function (when illuminated by a narrow beam) as does the translated strip (b).

The transformation requires changing the zone lens grating equation from rectilinear to polar coordinates. A decentered (off-axis) portion of the zone lens, as in (b) in Fig. 4.43, effectively includes a carrier frequency ν_o added to a symmetric zone lens, an average spatial frequency in the y direction. The rectilinear coordinate grating phase function of Eq. (1) of Section 4.4.3.1 is now written decentered,

$$\Phi_d(x, y) = 2\pi y\nu_o + \pi(x^2 + y^2)/f\lambda \tag{1}$$

$y\nu_o$ is the number of cycles subtended by y, changing phase as multiples of 2π. To transform to polar geometry, replace variables x and y by r and φ, yielding

$$\Phi(r, \varphi) = 2\pi r\nu_o + (r^2 + \overline{r_o}\varphi^2)/f\lambda \tag{2}$$

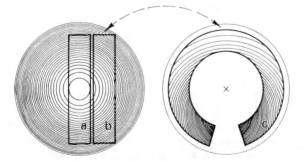

Figure 4.43. Illustration of zone lens (left side) showing sector strips for translational scan (a) on-axis and (b) off-axis. Strip (b) transformed to polar coordinates (right side) as annular band (c) for mounting on disk and rotation about its center x. Figure 4.37 illustrates their scanning configurations. (From Ref. Bry 1.) North Holland Physics Publishing Co., 1975.

in which r_o is the average radius of the hologram. The first term now represents a circular carrier.

Whether retained as a linear device or transformed as above, translation of a zone lens generates a corresponding displacement of its imaged focal point. Since the maximum displacement is the effective length S of the zone lens sector and the elemental displacement is equal to the spot size δ, the resolution is expressed simply as

$$N_s = \frac{S}{\delta} \tag{3}$$

the translational scan relation of Eq. (8) of Section 4.8.1. In the (quadratic function) zone plate, the spatial frequency ν is proportional to displacement. With $\nu = 1/d$ (d = grating spacing), then $\nu \approx \Theta/\lambda$ and the maximum scan half-angle $\Theta_{1/2}$ is $\Theta_{1/2} \approx S/2f$. This yields the maximum grating frequency (at the edge of the scan)

$$\nu_{max} = \frac{S}{2f\lambda} \tag{4}$$

The spot size δ is expressed as $\delta = a(f/D)\lambda$ (a = aperture shape factor for beam width D). Substituting this and Eq. (4) into Eq. (3) yields for $a = 1$,

$$N_s = 2D\nu_{max} \tag{5}$$

Since this translational motion actually generates (by Fig. 4.37) an angular scan $2\Theta_{1/2} = \Theta$, we can express it in angular form by substituting Eq. (4) into Eq. (5), yielding

$$N_s = \frac{\Theta D}{\lambda} \tag{6}$$

corresponding to Eq. (4) of Section 4.8.1 with the aperture shape factor $a = 1$. Figure 2.6 illustrates analogous operation by a conventional lenticular element, in which the angular deflection is given as $\alpha/2$ and the illuminating beam width is d. It is noteworthy that the factors which determine resolution of the CGH scanner are identical to those of any other, diffractive or conventional. In the case of the CGH scanner, Eq. (5) expresses resolution in terms of ν_{max}, the maximum grating frequency—a dominant parameter of CGH formation. It is, however, derivable from basic equation (6). While this scanner is mounted on a disk of active radius r_o and the diffracted wave forms a focal point a distance

f from the disk, this r_o is in quadrature with f. The projection of r_o on f is zero, hence the augmentation term of Eq. (19) of Section 2.8.3 is zero.

In Eq. (1) and (2) appear the term $f\lambda$ (focal length–wavelength), which means little to the computer plotter. These too can be transformed, with a substitution for the imaged spot size $\delta = (f/D)\lambda$. Thus $f\lambda = D\delta$ (for $a = 1$). Since the incremental step Δx of the translational scanner is $\Delta x \approx \delta$, then

$$f\lambda = D\Delta x = Dr_o\,\Delta\varphi \tag{7}$$

allowing Eqs. (1) and (2) to be expressed completely in instrumental terms.

A complication that arises in transforming a rectilinear coordinate zone lens to a polar one is the cross-coupling of the grating frequencies in the r and φ directions. That is,

$$\frac{\partial\Phi(r,\,\varphi)}{\partial\varphi} = 2\pi\nu_\varphi(r,\,\varphi) \propto r\varphi$$

tending toward spot aberration. To avoid this dependence (Bry 1), either make $r_o \gg D$ (larger disk or lower resolution for a given $\nu_{\max}$) or make an effective one-dimensional transformation (Bry 2), leaving focusing power only in x or φ. The second method requires following the scanner with a complementary cylindrical element to focus the scanned diffracted wave (which would otherwise form a line) to an isotropic point. Section 4.3.3 discusses a method based on one-dimensional illumination and restoration with cylinders (Ger 4). Other novel two-dimensional scan patterns were developed by Bryngdahl and Lee (Bry 2), introduced in Section 4.4.3.1. While feasibility was demonstrated by experiment, the resolutions were limited primarily by available plotting facilities, which could be enhanced significantly with laser beam or electron beam (Arn) recording. The resolution of the linear scanner was 300 spots (designed at $N = 320$). With a sufficiently high $\nu_{\max}$ and large D, spot resolutions in the thousands are possible. The more complex stitch-bar, triangular, and spiral scan resolutions were substantially lower, perhaps to 50 spots per scan. But the ingenuity of the coordinate transformations were clearly demonstrated. Subsequent work by Campbell and Sweeney (Cam) provides elegant consideration and experimental development of arbitrary scan patterns provided by CGHs operating in transmission at 633 nm and in reflection at 10.6-μm wavelengths.

4.5.4. Additional Considerations for Effective Operation

In addition to seeking to provide a scan pattern or wavelength of operation not readily available by conventional means, the use of computer-generated holograms for scanning is moderated by some distinctive factors. As derived from sharp-edged binary gratings, the diffracted output is rich in harmonic content, with the unwanted zeroth, −first, ±second, and ±third appearing most prominently. Although use of a sufficiently high carrier frequency allows us to filter out the undesired ones, even in phase-only (nonabsorptive) binary gratings, diffraction efficiency suffers from depletion of energy into the unwanted orders. A potential technique for restoring reasonable efficiency is to rerecord the CGH interferometrically in a thick hologram. Bartelt and Case (Bar 1) showed restoration of diffraction efficiency in the selected first order approaching 50%. Also, use of electron beam lithography recording full-scale on flat chrome-on-glass substrates shows potential for substantial diffraction efficiency (Arn) beyond the measured 40%. Another basic consideration is the attendant spot aberration that results from generating a grating pattern which scans an arbitrary spatial function, that is, one not formed by the rectilinear quadratic phase relationship represented in Eq. (1) of Section 4.4.3.1. This is discussed in Section 4.3.3.2.2 and is implicit in the precaution expressed earlier in transforming a rectangular coordinate zone lens to polar coordinates. See also Section 4.3 for a discussion of many related characteristics and processes.

5

Mechanical-Optical Integrity

5.1. ANAMORPHIC CROSS-SCAN ERROR REDUCTION

5.1.1. Fundamental Considerations

In the introductory paragraphs on scanned resolution (Section 2.8.1) were identified several terms that reappear here. They are $N_\perp$, the cross-scan resolution and the corresponding parameters $\delta_\perp$ and $w_\perp$, the cross-scan spot size and (raster) line-to-line center spacing, respectively. As expressed there, the along-scan resolution N is measured by the number of elements subtended per full active scan *in the direction of scan*. When a system is dedicated to providing along-scan only (cross-scan derived elsewhere or from a different portion of the system), it is assumed that $N_\perp = 0$; that is, the *along*-scan system executes no variability of placement of the focused spot $\delta_\perp$ in the *cross*-scan direction. This assures—as is typically required—that $w_\perp$ = constant, that is, no perturbation of line-to-line placements. When the system is to provide line scan only (no raster), the equivalent requirement is that the scanned lines superpose precisely.

Spot position error is most often generated by *angular* misplacement of the output beam. Concomitant with angular error may be translational error, in which the beam misorientation is accompanied by, or even dwarfed by linear misplacement. As is typical of many scanning systems, *when the output beam is collimated, its subsequent focusing results in the nulling of the translational error* at any scan angle position. The *angular error remains* and is often insidious in the cross-scan direction because of the resulting (very perceptible) unequal line spacings of a raster structure.

To deal with the error components—angular or translational, from focused or collimated beams—we invoke the awareness of scanned resolution of Section 2.8. Resolution can be considered in the form normally represented as "desired" or in the form associated with errors as "undesired." Regardless of the sources of angular or translational perturbations on a scanning system, their consequences in the imaged field are the same as those of purposeful action. Having identified $N_\perp$ as the cross-scan resolution, we retain that notation in discussing the undesired consequences of beam misorientation, with the understanding that they occur just as well in the along-scan direction. However, along-scan (random or pseudorandom) errors are subject to entirely different control techniques than those discussed here, for they are so intimately bound with the process that forms the "desired" along-scan resolution. That control typically entails direct electromechanical intervention and synchronization. Further, they can often be tolerated to a greater degree than cross-scan errors, for the perturbation is seldom upon adjacent (or even nearby) elements in the same scan line.

A principal equation [Eq. (19)] of Section 2.8 may now be expressed in cross-scan error form:

$$\Delta N_\perp = \frac{D_\perp}{a\lambda}\left(1 + \frac{r_\perp}{f_\perp}\right)\Delta\Theta_\perp \qquad \text{resolution elements} \qquad (1)$$

in which $D_\perp$, $r_\perp$, and $f_\perp$ are the cross-scan aperture size, radius, and focal length, respectively, and $\Delta\Theta_\perp$ is the angular misplacement of the deflected beam in the cross-scan direction. Figure 5.1 illustrates these parameters for a finite and positive $f_\perp$ and might be recognized as Fig. 2.5 in quadrature. The a and λ represent the aperture shape factor and wavelength, respectively, in which a seldom differs significantly from its along-scan counterpart. The $\Delta\Theta_\perp$ *is the problem*, and its consequence is $N_\perp$, measured in (usually fractional) units of resolution elements. For example, if one seeks the conditions for one-element spot misplacement, $\Delta n = 1$ (for $r = 0$ or $f = \infty$; subsequently discussed), then

$$\Delta\Theta_\perp = \frac{a\lambda}{D_\perp} \qquad (2)$$

recognized as the diffractive spread from the aperture $D_\perp$. That is, this error of

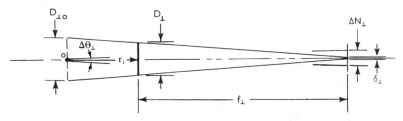

Figure 5.1. Parameters associated with cross-scan errors, in quadrature with desired along-scan performance of Figure 2.5. Angular error $\Delta\Theta_\perp$ causes focused spot $\delta_\perp$ misplacement of $\Delta N_\perp$. Focal distance $f_\perp$ shown finite and positive.

$\Delta \Theta_{\perp}$ misplaces the beam by one (diffraction-limited) resolution element in the cross-scan direction. Numerical evaluation provides rapid appreciation for the magnitude of the angular errors which may be tolerated, especially when measured in fractional units. Also, following Section 2.8.4, mechanical error $\Delta \Phi$ can be magnified by a factor $m_{\perp}$, where $m_{\perp} = \Delta \Theta_{\perp} / \Delta \Phi_{\perp} \geq 0$, such that [per Eq. (28b) in Section 2.8.4] Eq. (1) is now represented more completely as

$$\Delta N_{\perp} = \frac{D_{\perp}}{a\lambda} \left(m_{\perp} + \frac{r_{\perp}}{f_{\perp}} \right) \Delta \Phi_{\perp} \qquad (3)$$

Transmission-type holographic scanners with wobble correction (Section 4.2.2.4) effectively operate with $m_{\perp} \rightarrow 0$ and $f_{\perp} = \infty$.

5.1.2. Anamorphic Beam Handling for the Deflecting Aperture

This technique is well recognized for use in conventional laser scanners; notably in those using polygons having excessive variation in facet-to-axis angle (cone ⟡ error) or bearing wobble. It can also provide effective reduction of cross-axis angular errors in holographic scanning systems. While many of the characteristics of design and operation are similar in conventional and holographic systems, certain differences do exist which allow either greater flexibility or require special consideration. As in our presentation on scanned resolution in Section 2.8, a general approach is motivated here, too. While our main concern is with the use of anamorphics for holographic scanners, we benefit significantly from its commonality in conventional systems. Distinctions and special techniques will be highlighted after the basics are expressed.

Referring to Eq. (3), for finite $m_{\perp}$, the principal contributor to cross-scan perturbation is $D_{\perp}$, the cross-scan aperture size. If $D_{\perp} \rightarrow 0$, then $\Delta n_{\perp} \rightarrow 0$. This is the basis for most anamorphic techniques for reduction of cross-scan error. Since $\Delta N_{\perp}$ is directly proportional to $D_{\perp}$, a corresponding fractional correction results from incomplete reduction of $D_{\perp}$ to zero. Although not originally expressed in this manner, this fundamental principle describes the familiar polygon correction systems introduced by Fleischer (Fle) in 1973 and adapted to many variations, depending on the locations and form of the anamorphic elements. The functional manner by which this is accomplished is illustrated in Fig. 5.2 representing a generic scanner H operating in the transmission mode. Figure 5.2b shows the throughput beam coaxial for simplicity, for the diffracted angles of a holographic scanner do not alter this operation.

The plan view (Fig. 5.2a) illustrates the along-scan process, which is essentially unmodified by the addition of cylinders L_1 and L_2, oriented to contribute no optical power in the x direction. The lens L_o is a typical flat-field objec-

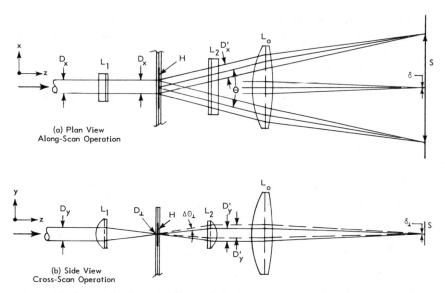

Figure 5.2. Anamorphic beam handling at deflecting aperture. Generic scanner H operating in transmission. Side view (b) shows throughput beam coaxial for simplicity.

tive which accepts the collimated light of aperture $D_x' = D_x$ which was deflected through angle Θ by facet H and focuses it to a spot of size δ on the image surface S. The side view (Fig. 5.2b) illustrates the cross-scan process: L_1 now exhibits power in the y direction and focuses D_y to $D_\perp$ on H, whereupon the undeviated (solid line) beam rediverges, is captured by L_2 such that $D_y' = D_y$, and continues to be focused by L_o to form spot size $\delta_\perp$ on the image surface S. L_1 and L_2 are separated by the sum of their focal lengths, intersecting at H. When a $\Delta\Theta_\perp$ cross-scan error occurs, the angularly displaced diverging beam (dashed lines) is again collected by L_2, recollimated to the translated (dashed lines), D_y' and refocused by L_o to the same $\delta_\perp$ on S. The angular error $\Delta\Theta_\perp$ at H is transformed by L_2 to tolerant translation of the collimated beam into L_o, maintaining a stationary $\delta_\perp$. While anamorphism of L_1 and L_2 is customarily introduced with single cylinders as shown (and L_2 may be toric to present more uniform optical power to the scanned beam through the angle Θ), they may be in the form of cylindrical (confocal telescope) or prismatic beam compressors (for L_1) and expanders (for L_2). Thus $D_\perp$ can be selected to exhibit any fraction of the original aperture D_y, whereupon Eq. (3) is operative for determining the resulting fractional spot misplacement. For $m = 1$ and $r = 0$, it reduces to

$$\Delta N_\perp = \frac{D_\perp}{a\lambda} \Delta\Theta_\perp \qquad (4)$$

such that the angular error is reduced by a factor,

$$R_\Theta = \frac{\Delta N_\perp}{\Delta N_y} \frac{D_\perp}{D_y} \tag{5}$$

To restore the resulting focused spot to quadrature equality, that is, $\delta = \delta_\perp$ (and to isotropy if intended as a round cross section), then the cross-scan aperture D_y' must at some later stage be restored to equal the along-scan aperture D_x', where the primed terms represent the dimensions at a location beyond the deflector. This may even occur beyond L_o or its equivalent. Herein lies the major freedom available in such cross-scan error handling, in that the $D_\perp$ at the deflector may be adjusted independently to reduce angular error. When restored at a subsequent stage, it allows equally free adjustment of the size of the resulting $\delta_\perp$. An interrelationship between these factors will be discussed in Section 5.1.3 when translational error is added to angular error. As in the synthesis of any information-handling system, $\delta_\perp$ need not necessarily be equal to δ, typified by the use of elliptical spots for various purposes, including enhanced detectivity in scanning generally unidirectional information and in varying overlap characteristics during raster formation.

A variation described in 1974 for holographic scanning (Bei 7) was expressed earlier in discussion surrounding Fig. 4.19. There is shown the use of a single anamorph, corresponding to L_2. The conjugate holographic exposure (back through the same cylinder, effectively L_1) provides the requisite aperture reduction to $D_\perp$. Not shown in that view, the overilluminated along-scan aperture D is substantially larger than $D_\perp$. Hence, for isotropic focal point, the error reduction ratio is $D_\perp / D$ (for $m_\perp = 1$ and $f_\perp \gg r_\perp$).

5.1.2.1. Exemplary Technique: Illumination Normal to Holographic Disk

Anamorphic beam handling not only reduces cross-scan errors due to wobble, but suppresses all angular deviations in the narrowed-aperture direction. A more recent variation for holographic scanning by Brasier (Bra 3) merits attention because of its novel and integrated utility, directed mainly toward straightening the arcuate scan from a radially symmetric holographic scanner (see Sections 4.1.4 and 4.2.2.2). Also provided is a unique manner of gaining a high duty cycle in underilluminated operation while using a surprisingly small holographic disk.

Figure 5.3 is an annotated form of the referenced illustration. The expanded beam 16 from laser beam 32 (through lenses 34 and 38) is directed on first cylinder 54 which compresses it (rather conventionally along 56 in the y direction) to line 58 on holographic disk 42. Observe, however, that the (plane linear) grating lines on disk 42 are oriented such that the central line of each facet

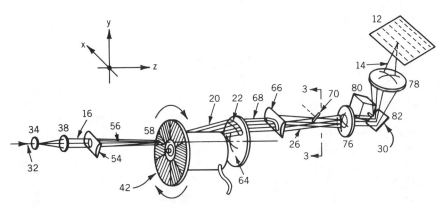

Figure 5.3. Anamorphic beam handling which straightens an arcuate scan, reduces cross-scan wobble, and provides increased duty cycle (see the text). (Adapted from Ref. Bra 3.)

is *radial* (and parallel to line 58 when the facet is centered). This is in quadrature to the more typical grating orientation, as represented schematically in Figs. 4.2 and 4.3. The scanned output beam 20 remains an arcuate locus 22, but because of the quadrature orientation, the locus is generally radial to the disk (rather than tangential to it) and directed generally along x in this coordinate system.

Intermediate objective lens 22, focused on the center of line 58, transforms the diverging y component to collimated at 68 and also transforms the scanned collimated x component at 20 to focus in the plane of section 3-3. The second cylinder 66 operates only on the collimated y component at 68 to focus it via 26, also in the plane of section 3-3. In this intermediate image plane, the arcuate scan of beam 20 is straightened along x as an image of straight line 58, satisfying one principal purpose of this implementation. The image in the plane of section 3-3 is also stabilized against wobble in the y direction. Telecentric lens 76 retransforms the x-scanned beam to a stationary aperture at 30, so that galvanometer 80 mirror 82 can impart the desired cross-scan y component to form a raster in plane 12 via flat-field lens 78. (Mirror 70 is a pick-off for x-scan synchronization.)

Thus far the entire process could have been conducted with the more typical grating orientation of Figs. 4.2 and 4.3. This quadrature arrangement provides, however, a unique virtue not readily available from mirror deflection systems—dramatic improvement of duty cycle η_c which is directly proportional to optical throughput efficiency and to resolution. By Section 2.8.1 (for $\dot{\Theta}$ = constant), $\eta_c = \Theta / \Theta_{max}$, the ratio of the useful scanned angle to that total available, and by Section 2.8.4, is independent of a constant magnification m. In this under-illuminated mode, $\eta_c = 1 - \Delta$, where $\Delta = \tau / T = D_\perp / D_{max}$. For example, when $D_\perp = 0.5 D_{max}$ (as in some conventional systems), $\eta_c = 0.5$ and when $D_\perp = 0.05 D_{max}$ ($\tau = 0.05 T$), then $\eta_c = 0.95$—almost double the duty cycle with no increase in scanner size (for same number of facets). This original implementation utilized a 10-faceted scanner disk $2\frac{3}{8}$ in. in diameter.

5.1.3. Translational Error: Output Beam Nonnormal to Diffractive Surface

Having concentrated on the angular errors, for they are most burdensome and costly to constrain by direct mechanical integrity, an interrelated translational error can develop under certain conditions, requiring distinctly different control. Translational error develops in a holographic scanner when the input and output beams are nonnormal to the grating surface, as represented in Fig. 4.30A. Depending on the location of the fulcrum of the error $\Delta\alpha$, the principal output ray will be derived from a different point on H. This can be measured by a *translational* error Δh at H, executed by and normal to the principal output ray. If the output beam is collimated, no problem. When focused, however (per Fig. 5.2b) and H is nonnormal to the beam and rotated through $\Delta\alpha$, the aperture $D_\perp$ is displaced on H (in the y direction), causing a corresponding displacement Δs of its image $\delta_\perp$ on surface S. Now we have an interrelationship between the angular and translational error. By Eq. (5), $D_\perp$ determines the angular error reduction ratio $R_\Theta = D_\perp / D_y$, while the image spot size $\delta_\perp$ is established by the magnification $M_\perp$ between $D_\perp$ and $\delta_\perp$,

$$M_\perp = \frac{\Delta s}{\Delta h} = \frac{\delta_\perp}{D_\perp} \qquad (6)$$

leading to an error reduction product,

$$R_\Theta M_\perp = \frac{\delta_\perp}{D_y} = c \qquad (7)$$

where the lower the value of either R_Θ or $M_\perp$ provides improvement for angular and translational errors, respectively. There now appears a conflict. If one seeks to provide drastic reduction in angular error by making $D_\perp$ very narrow per Eq. (5), then translational error is in jeopardy, by Eq. (6). The ratio $\delta_\perp / D_y$ in Eq. (7) is a constant, for $\delta_\perp = k (f/D_y) \lambda$, a design criterion. $\delta_\perp$ and D_y are not normally adjustable independently. The next section discusses a special case.

This balanced condition was handled effectively in a design by Funato (Fun 3). A demagnification $M_\perp$ of about $\frac{1}{10}$ was reserved for reduction of translational error by that factor, while providing significant (40:1) improvement in line straightness from a disk scanner having $\Theta_i \approx \Theta_o \approx 44°$ with added anamorphic optics. Although not expressed explicitly, the extra demagnification of $M_\perp \approx \frac{1}{10}$ was facilitated by positioning the second cylinder L_2 of Fig. 5.2b close to the image plane S. It must be emphasized that had a focusing anamorphic system not been employed for scan-line straightening, this translational problem, meriting a commendable solution, would have not developed.

An attempt to control near-collimated beam translation was made with the use of, effectively, an objective lens split into two parts (Dog). The first part is a cylinder oriented with power in the along-scan direction before the deflector followed by the second part, which is a cylinder with power in the cross-scan direction after the deflector. The cylinders, in quadrature, are focused on the same image surface, forming effectively a "thick" objective lens. While the second cylinder will reduce translational errors (if they existed), there appears no benefit (along with complications in complexity and image quality) over replacement of both cylinders with a conventional objective lens following the deflector. As emphasized earlier, *translation of collimated light on an objective lens will not misplace the focal point.*

5.1.4. Anamorphic Beam at Error Fulcrum

Returning to Eq. (3), we observe that there are *two* sets of variables that allow $\Delta N \to 0$. The first, discussed exclusively thus far, is $D_\perp \to 0$, which yields the error reduction ratio $D_\perp / D$ for a scanned isotropic focal point. The second set of variables separable from Eq. (3) is the parenthetic one equated to zero,

$$m_\perp + \frac{r_\perp}{f_\perp} = 0 \qquad (8)$$

yielding

$$f_\perp = -\frac{r_\perp}{m_\perp} \qquad (9)$$

where these cross-scan values are resolved collinearly, as exemplified in Fig. 5.1.

This intriguing result suggests that the cross-scan error may be nulled *independently of the real value of* $D_\perp$. In Fig. 5.1, where $m = 1$, the scanned output beam must *diverge* from a virtual vocal point o coincident with that of $r_\perp$, as represented in Fig. 5.4. As long as the error $\Delta \Theta_\perp$ propagates from the same center o, it does not affect the image in the cross-scan direction. The cylindrical lens $L_\perp$ focused on o will transform the angular error to a translating collimated beam, which when focused by an objective lens results in a single image point. Systems more adaptable to such operation are those having surface components

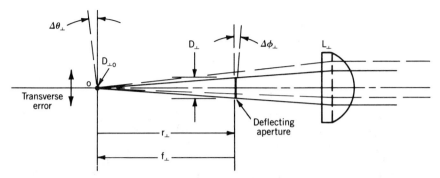

Figure 5.4. Anamorphic beam handling at the error fulcrum. Error angle $\Delta\Theta_\perp$ about center o propagates (dashed lines) through aperture $D_\perp$ and is nulled at cylindrical lens $L_\perp$. Aperture $D_\perp$ effectively shifts (not shown) with dashed lines during angular error. Example of $m = 1$, whence $f_\perp = -r_\perp$. (See the text.)

normal to the rotating axis, described in Section 4.2.1 and in which the deflecting element is displaced a finite distance r from the rotating axis. This novel technique, to the knowledge of the author, has not yet been applied.

The basis for its action derives from the augmentation process introduced in Section 2.8.2, in which Fig. 2.5 illustrates increased resolution (positive augmentation) due to the effective aperture D_o at the fulcrum being greater than the real aperture D. In negative augmentation, the effective aperture is smaller than the real one. In this case the effective aperture $D_{\perp o}$ is zero, hence no cross-scan error. Note that for $m \neq 1$, the fulcrum o moves to $r_{o\perp} = r_\perp / m_\perp$ per Section 2.8.4 and Fig. 2.8. This is reflected in Eq. (9) for the solved value of $f_\perp$.

A limitation to this form of correction is that it depends on the stationarity of point o (of Fig. 5.4) during the cross-scan error excursions. If the error only pivots about o, it operates effectively. If, however, $D_{\perp o}$ also executes a transverse component (identified as "transverse error" in Fig. 5.4), that component will not be removed. Such transverse error will develop, for example, when the aperture $D_\perp$ executes a coning angular displacement $\Delta\Phi_\perp$ and the error fulcrum moves from o to the surface of D. Thus it is not as general a correction process as that of $D_\perp$ reduction, which operates to rectify this source of cross-scan error.

5.2. SCAN LINEARITY

5.2.1. Introduction

Scan linearity requires a constant spot velocity v_x (along-scan) on the image surface, derived from a uniform angular velocity of the rotating deflector $\dot{\Phi}$, as introduced in Section 4.1.1. Ultimately, geometric spacing of equally spaced ideal data points need be uniform. When $v_x/\dot\Phi$ is not constant, this poses a special problem for analog signals. When digitized, pixel spacing may be rectified by electronic adjustment of pixel timing, derived from an open-loop error function, or closed-loop from, for example, scanning a pilot beam across a reference linear grating (Toy). This section addresses the basic optomechanical scan linearity to approach $v_x/\dot\Phi$ = constant over a useful operating range.

Although some systems, such as bar code scanners, can allow significant nonlinearities ("distortion" in optical terminology), others, such as image recorders, are very intolerant to such geometrical errors. Because nonlinearities are typically slowly varying, the electronic data logic of some systems can compensate for significant nonlinearities. Business graphics recorders and printers typically limit nonlinearity to about 1%, while graphic arts and analytic instruments are limited to from 0.1 to 0.01% nonlinearity, where 0.01% of, say, 10,000 elements per half-image of 20,000 elements per scan misplaces the image point by 1 pixel from nominal.

Near-perfection in scanned linearity is achievable uniquely with a radially symmetric system in which the image-bearing surface also conforms concentrically. See, for example, Fig. 4.2 and related discussion on radial symmetry in Section 4.1.2. Indeed, the early superresolution holographic systems (see, e.g., Figs. 4.10, 4.16, and 4.20) and even some of the lower resolution ones (by Figs. 4.23 and 4.27) maintained radial image symmetry to eliminate flat-fielding optics while providing high placement integrity of the image points.

For flat-fielded systems that can tolerate some (0.1 to 1%) nonlinearity, one may generate an intrinsically nonlinear function and correct for it by control of the hologram or subsequent optics, or both. By far the easiest way to accomplish this is to maintain normally directed radial symmetry and to add "f-θ" flat-field optics typical of that developed over many years of nonholographic scanning activity. This approach is exemplified in Figs. 4.14 and 4.22 and is implicit in the work represented by Fig. 4.28. When departure from radial symmetry is implemented, with or without normal output beam directions, heroic means are necessary to achieve geometric linearities of such quality. These departures may be taken to allow flat disk substrate utilization and can include near-Bragg operation to reduce wobble sensitivity. Along with the concomitant scan-line bow and the approach to its straightening discussed in Sections 4.2.2.5, 4.2.2.6.3, and 4.2.2.6.4, there is an inevitable scan nonlinearity. Correction for this distortion is the subject discussed here.

5.2.2. Disk Scanner Nonlinearity

5.2.2.1. Counterrotating Gratings

The earliest investigation of disk scanner nonlinearity was published by Wyant in 1975 (Wya) and discussed in Section 4.2.2.5.2. While the configuration of counterrotating plane gratings is unique and not directly adaptable to the more familiar single grating scanners, the conceptual approach is noteworthy. In that design the scanned output angle Θ of a collimated beam varies as the sine of the grating rotation angle Φ, relatively rapid and linear in the center and compressed to extreme nonlinearity at the ends. To compensate for this, Wyant invoked the use of an idealized focusing lens which retains the transfer function $x = f \tan \Theta$ ($x =$ displacement on image plane and $f =$ focal length). Thus its tangent expansion toward the margins of scan tend to complement the foregoing sinusoidal compression in that region, reducing distortion significantly over a useful operating range.

No indication is provided of realizability of the lens having this ideal function when it is also required to provide a significant "pupil relief" distance—space from the scan nodal point to lens mount surface. Also, while some clever folded configurations were presented, this lens would need to accommodate beam wander over a large aperture and operate effectively from two different nodal points. Notwithstanding these requirements, which may be resolved, the concept of burdening the lens with a complementary error function is useful and general.

5.2.2.2. Near-Bragg Condition Disk Scanner Nonlinearity

One of the most important nonradially symmetric configurations is that of the disk scanner operating near the Bragg regime. The wobble and line bow characteristics of this scanner have been covered earlier in Sections 4.2.2.4 and 4.2.2.5, while several variations providing convergent beam output were discussed in Section 4.2.2.6. The first requirement toward control of scanned linearity is to understand its functional characteristics. Specifically, given its operational parameters, the output scan function versus disk rotation will reveal information that can lead toward corrective procedure.

A development due to Kramer (Kra 6) provides a relationship, expressing scanned displacement x on the image plane. As adapted here,

$$x = \frac{zk \cos \Theta_o \sin \Phi}{\left(\cos^2 \Theta_i + 2k \sin \Theta_i \cos \Phi - k^2\right)^{1/2}} \qquad (1)$$

in which z is the distance from scanner to image plane and $k = \lambda/d$. Applying some characteristics of typical operation, in the vicinity of $\Theta_i \approx \Theta_o \approx 45°$, where from the grating equation $k = \sqrt{2}$, this reduces to

$$x(45°) = \frac{z \sin \Phi}{\left(2 \cos \Phi - 1.5\right)^{1/2}} \qquad (2)$$

and in the vicinity of $\Phi = 0$ (center of scan),

$$x(45°, \Phi = 0) \approx z(\sqrt{2}\ \Phi) = z\Theta \tag{3}$$

The factor $\sqrt{2}$ will be recognized as the magnification $m = \Theta/\Phi$, by defining equation (7) of Section 2.8.1, taken at the center of scan. It reappears in Sections 2.8.4 and 4.2.2.5.1.

To evaluate the scan function under typical operation represented by Eq. (2), we compare it to a nominal one represented by

$$x_n = z \tan \Phi \tag{4}$$

This is the scanned point displacement on an image plane which is mounted normal to the principal ray when $\Phi = 0$. The ratio of actual/nominal scan is the ratio of Eq. (2) to Eq. (4),

$$m_n = \frac{\cos \Phi}{(2 \cos \Phi - 1.5)^{1/2}} \tag{5}$$

Again, at $\Phi = 0$,

$$m_n(\Phi = 0) = \sqrt{2} \tag{6}$$

Equation (5) is the scan magnification at any Φ, while, Eq. (6) is the magnification at the center of scan. If the magnification were maintained constant at the same value as at $\Phi = 0$, the scan would be linear. The actual linearity is therefore the ratio of Eq. (5) to $\sqrt{2}$, designated as r_n:

$$r_n = \frac{\cos \Phi}{(4 \cos \Phi - 3)^{1/2}} \tag{7}$$

$\Theta = m_n\Phi$ represents disk scanner performance in the vicinity of $\Theta_i = \Theta_o = 45°$, plotted in Fig. 5.5 over a maximum practical hologram rotation angle Φ. For Φ greater than $25°$, not only does nonlinearity become extreme, but scan bow becomes untenable. The plot shows the actual scan angle Θ compared to the ideal when $m = \sqrt{2}$ (Θ scale on left side) and also provides the curve for r_n (scale on right side), both versus the rotation angle Φ. Indices of performance as follows: At $\Phi = 5°$, $10°$, $15°$, $20°$, and $25°$, the ratio $r_n = 1.0034$, 1.016, 1.039, 1.079, and 1.146, corresponding to nonlinearities of 0.34, 1.6, 3.9, 7.9, and 14.6%, respectively. This is the information we seek. For example, if a system can tolerate up to 3.9% nonlinearity, scanner rotation of $\Phi = \pm15°$

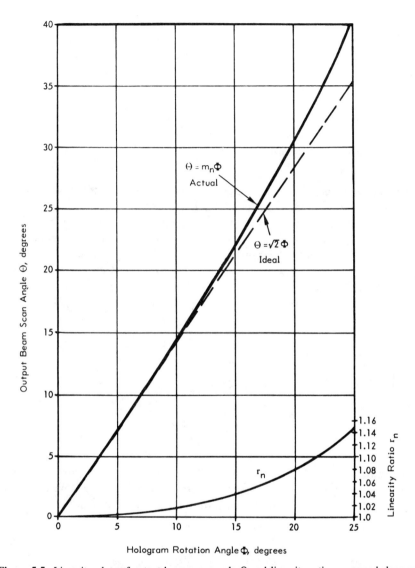

Figure 5.5. Linearity plots of output beam scan angle Θ and linearity ratio r_n versus hologram rotation angle Φ for transmission-type disk scanner having $\Theta_i = \Theta_o = 45°$. Actual and ideal scan angles are compared and their ratio is represented by r_n.

would provide a half-scan angle of $15°(1.039)$ ($\sqrt{2}$) $\approx 22°$ ($\Theta \approx 44°$ full scan angle). This assumes, of course, either a focusing system such as that described in Section 4.2.2.6.4, or a preobjective system launching a collimated beam into a perfect $f\text{-}\Theta$ lens for flat fielding. Section 5.2.3 provides additional discussion of this characteristic. The focusing system (Iwa 2) was limited to $\Phi = 15°$ to provide reasonable freedom from bow, but did not address scan linearity as interpreted here.

5.2.2.3.   Scan Magnification Near Center of Deflection and General Nonlinearity

The work discussed above considered scan magnification in the vicinity of $\Theta_i = \Theta_o = 45°$ over a range of rotation angle Φ. Another important operational condition is the scan magnification for arbitrary Θ_i and Θ_o in the vicinity of $\Phi = 0$ (close to the center of deflection), where the value of scan magnification is dominant. This can be determined by following a similar procedure, taking the ratio of the general equation for x deflection [Eq. (1)] to the nominal one [Eq. (4)], yielding

$$m_x = \frac{k \cos \Theta_o \cos \Phi}{\left(\cos^2 \Theta_i + 2k \sin \Theta_i \cos \Phi - k^2\right)^{1/2}} \qquad (8)$$

At the center of scan, $\Phi = 0$, whereupon

$$m_o = \left[\frac{\cos \Theta_o}{\left(\cos^2 \Theta_i + 2k \sin \Theta_i - k^2\right)^{1/2}}\right] k \qquad (9)$$

Substituting for $k = \lambda/d = \sin \Theta_i + \sin \Theta_o$ (the grating equation) and performing some operations, the bracketed term reduces to unity, revealing the rather surprising simplicity of the magnification at center of scan,

$$m_o - k = \frac{\lambda}{d} \qquad (10)$$

This was developed heuristically at the close of Section 2.8.4 and is used to estimate the central scan magnification of the IBM POS scanner (having $\Theta_i = 22°$ and $\Theta_o = 45°$) in Section 4.2.2.6.1.1.

A general expression for scan linearity is obtained by dividing the general equation (8) for scan magnification by the factor k. That is,

$$r_x = \frac{m_x}{k} = \frac{\cos \Theta_o \cos \Phi}{\left(\cos^2 \Theta_i + 2k \sin \Theta_i \cos \Phi - k^2\right)^{1/2}} \tag{11}$$

Equation (7) for r_n is a special case, when $\Theta_i = \Theta_o = 45°$.

5.2.3. Disk Scanner Linearization

The developments discussed above provide requisite information for scan linearization. The case illustrated at the end of Section 5.2.2.2, having $\Theta_i = \Theta_o = 45°$ and $\Phi = \pm 15°$, reveals 3.9% nonlinearity. The functional relationships are expressed by Eqs. (5) and (7) and illustrated in Fig. 5.5. If the scanned beam output is focused directly on a flat image surface without interposing additional optics (Iwa 2), one can expect to measure a corresponding nonlinearity.

With additional optics—and, in particular with a modified flat-field lens, this amount of nonlinearity can be reduced significantly. An ideal lens provides the image function $x = \tan \Theta$. A corrected flat-field lens is purposefully adjusted to overcome the scanned expansion with tan Θ—thus identified as an f-Θ lens—utilized widely in conventional scanners having uniform magnification. The disk scanner nonlinearity is shown to *accentuate* this expansion, exemplified by the plot of r_n in Fig. 5.5. It is required, therefore, to overcompensate the lens; that is, to impart additional complementary barrel distortion, requiring skillful manipulation of lens characteristics (Lev) while satisfying all regular requirements for spot quality, field flatness, and uniformity.

An example of scan lens design (Her) addresses this type of disk configuration at $\Theta_i = \Theta_o = 45°$ but does not account for the additional nonlinearity developed here. It expresses the scan magnification as uniform and equal to $A = \sin \Theta / \sin \Phi$ rather than $m = \Theta / \Phi$. There is, of course, a close correspondence between the two functions over the limited range of Φ. Mainly, however, this work only compensated for the nominal tan Θ function of an "undistorted" lens, and yielded one having characteristics very similar to the conventional f-Θ lens.

A uniquely different design is noteworthy, although its original objective was to emulate the characteristics of a conventional glass f-Θ flat field lens with a holographic lens (Ono 5). With an extension to their generalized zone lens work discussed in Sections 3.4.3, 4.2.2.6.2, and 4.2.2.6.3, Ono and Nishida succeeded quite effectively. The reported design provides excellent linearity and focusing characteristics when applied as a flat-field lens for a conventional polygon scanner. With the flexibility available from astute manipulation of the generalized zone lens, the possibility exists of providing some additional barrel compression to complement the expansion of the disk scanner while retaining good flat-field characteristics.

5.3. SUBSTRATE ANGULAR AND TRANSLATIONAL ERRORS

5.3.1. Sensitivity to Wobble and Translation

Wobble sensitivity was expressed in Section 4.2.2.4 during discussion of operation in the Bragg regime. An analogy between transmissive Bragg diffraction and prism operation in refraction was presented, yielding corresponding error null conditions. When the Bragg condition is approached, the output ray group will be affected minimally by angular error of the substrate, considered then to be a flat disk. In Section 4.2.1.3 we discussed transmissive scanners having surface component normal to the axis, and in Section 4.2.1.3.3 we introduced consideration of wobble sensitivity of the conic substrate scanner. Because this may be compounded with a translational error, a more complete interpretation is developed here, starting with the transmissive conic substrate and then generalizing to any substrate operating in transmission or reflection.

Tipping errors manifest differently from two conditions of beam output characteristic:

1. That developed in Section 4.2.2.4. This relates to tipping of the substrate within fields of illuminating and diffracting *collimated* light, where the output beam angle is unaffected by substrate *translation*.
2. Nutation of the substrate about the axis and diffracting a *converging* beam to focus. This develops two distinct components of focal-point misplacement in the image surface:
   a. Due to beam *angle* error of (1) propagating over the focal length
   b. Due to misplacement of the holographic lenticule with *translation* of the substrate; hence the position of its focal point

Consider a conic substrate of active radius r which launches a converging beam over a distance f to its focal point p (see Fig. 5.6). The substrate is tipped erroneously about its radial center through an angle $\Delta\alpha$, causing the output beam principal ray to be angularly displaced through $\Delta\Theta_o$. The displacement of focal point p to point p' on a normal image surface is the sum of the angular and translational errors

$$\Delta h = \Delta h_a + \Delta h_b \qquad (1)$$

$$= f\Delta\Theta_o + r\Delta\alpha \qquad (1a)$$

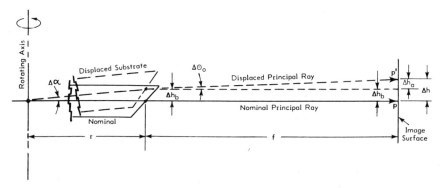

Figure 5.6. Angular and translational errors in lenticular holographic scanner due to substrate angular displacement $\Delta\alpha$ causing linear displacement Δh_b and angular error $\Delta\Theta_o$. Nominal focal point p is displaced through Δh to point p'. Substrate radius is r and its hologram focuses over distance f.

in which the first term is due to (a) above and the second due to (b). This is independent of the illuminating beam subtense and is basically similar to the resolution considerations of Section 2.8.

From Eq. (28) of Section 4.2.2.4, which determines the angular error component $\Delta\Theta_o$, the total displacement becomes

$$\Delta h = \left\{ r + f\left[1 - \frac{\cos(\Theta_i + \Delta\alpha)}{\cos(\Theta_o - \Delta\alpha)} \right] \right\} \Delta\alpha \qquad (2)$$

A development by Funato (Fun 1, Fun 2) yields the following representation for the total misplacement of the focal point, in corresponding nomenclature:

$$\Delta h_f = \left\{ r + f\left[\frac{\cos\Theta_o - \cos\Theta_i + (\Delta\alpha/2)\sin\Theta_i}{\cos^3\Theta_o} \right] \right\} \Delta\alpha \qquad (3)$$

The principal distinction is the $\cos^3$ term in the denominator which develops from multiplication of $\cos\Theta_o$ by an extra $\cos^2\Theta_o$ term, based on a different assumption for Δh_a. The $(\Delta\alpha/2)\sin\Theta_i$ term in the numerator develops from a Taylor series expansion of the displaced grating equation [similar to that of Eq. (24) of Section 4.2.2.4] and is even smaller than the already small $\Delta\alpha$.

Thus, using the appropriate first power term in the denominator and neglecting the terms in $\Delta\alpha$, both Eqs. (2) and (3) reduce to the simplified form

$$\Delta h = \left\{ r + f\left[1 - \frac{\cos\Theta_i}{\cos\Theta_o} \right] \right\} \Delta\alpha \qquad (4)$$

These relations need not be limited to a conic section substrate. They apply to any substrate having a component of its diffracting surface normal to the rotating axis and having an output beam focused along f also generally normal to the axis (in the r direction). They can be generalized to apply to any output beam oriented at an angle Θ_r to the r direction (where r is the distance to the error-nutation center) by providing for a $\cos \Theta_r$ variation upon r in Eq. (1a). Following Eq. (2) here and Eq. (29) of Section 4.2.2.4, this leads to the general expression

$$\Delta h = \left\{ r \cos \Theta_r + f \left[1 \mp \frac{\cos (\Theta_i + \Delta\alpha)}{\cos (\Theta_o \mp \Delta\alpha)} \right] \right\} \Delta\alpha \qquad (5)$$

where the upper sign applies for transmission and the lower for reflection gratings, and Θ_r is the angle between the output beam and the vector direction of r. The image surface is assumed normal to the output beam. In reflection, tipping errors are aggravated. A method proposed by Kramer (Kra 8) utilizes double reflection (as by a corner reflector, adapted to some conventional scanners) to reilluminate the diffracting facet such that the error tends toward null.

5.3.2. Sensitivity to Surface Deformation

5.3.2.1. *Characteristics and Tolerance Criteria*

Surface deformation can be considered as a localized form of substrate wobble discussed in Section 5.3.1. Indeed, the principal ingredients for analysis of surface deformations are contained in prior discussions of gross effects of misalignment of parallel surface diffractors and prismatic refractors. Distinctions arise in interpretation of the parameters and allocation of magnitudes to the degree of tolerable surface deformation.

Deformations of supporting substrates may appear in several forms. Consider a reflective hologram at which diffraction occurs at the outside surface. Clearly, that is the only surface of interest. If, however, the hologram is on the inside surface of a protective transparent member, two surfaces must be considered. Taking the outside surface reflective case first for simplicity, the deformation of interest is one whose subtense is in the order of magnitude of the size of the optical aperture D being illuminated, forming a spot size δ of dimension $\delta = a(f/D)\lambda$, where a is the aperture shape factor and f is the focal length from the focusing element to the image surface. Any deformation of size less than $0.1D$ may be considered more like scatter, as long as the average surface integrity is high. Similarly, any surface deformation of size greater than $10D$ falls more into the wobble domain discussed earlier—again, as long as the average quality around D is acceptable. Interest here is in that vicinity around D.

Two problems can arise with surface deformation:

1. Aberration of spot size due to misdirected ray trajectories from within the aperture.
2. Spot centroid displacement due to asymmetric bias of ray trajectories. In the limit, the spot size δ may remain integral but be displaced.

Either or both can develop, depending on the character of the deformation and the changing relationship of illuminated aperture D to the substrate during scan. To quantify tolerable magnitudes of deformation, we can combine both effects by considering that the aperture at one instant is correct and at another instant is misplaced through a surface angle error $\Delta\alpha$ which generates ray displacement through $\Delta\Theta$. For case 1, the two instants are effectively simultaneous due to the two different portions of the aperture exhibiting different surface orientations. For case 2, the two instants are separated sufficiently to form two successive spots δ separated through an error angle $\Delta\Theta$ (where, if there were no deformation, they would either superimpose or be separated through a required displacement).

The tolerable spot aberration or misplacement varies significantly depending on the acuity of the application. For the sake of problem illustration, assume that the correct spot of size δ is paired with an incorrectly positioned one having a maximum displacement of 20% of its FWHM (full width at half maximum). In case 1, these occur simultaneously, widening the spot in the direction of error. In case 2, these errors occur successively, misplacing the spot by that factor. The 20% of FWHM criterion is not atypical, but specific tasks require detailed assessment of that tolerance, which could be more stringent by at least an order of magnitude for continuous tone images (Sch). Assume a nonvignetted gaussian aperture distribution, for which the aperture shape factor to $1/e^2$ of δ is $a = 4/\pi$ and the ratio of FWHM to $1/e^2$ widths is 0.589; then the a at FWHM is $(4/\pi)0.589 = 0.75$. Thus $\delta(\text{FWHM}) = 0.75\,(f/D)\,\lambda$. A 20% criterion allows a displacement $\Delta = 0.15(f/D)\,\lambda$, yielding the tolerable angular error of

$$\Delta\Theta_m = \frac{\Delta}{f} = 0.15\,\frac{\lambda}{D} \qquad (6)$$

Beam misplacement angles $\Delta\Theta_m$ result from surface deformation angles. Our task is to identify them and clarify their tolerable effects.

5.3.2.2. Surface Deformation in Reflection

In holographic reflection, when the diffractor is on the outside, only one surface may be considered. It is intended nominally to be at the correct angle. If not, it suffers wobble. Thus the average error is here assumed zero. The problem reduces to determining the localized error $\Delta\alpha$ allowed on that surface which generates the tolerable angular displacement $\Delta\Theta_m$. From Eq. (29) of Section 4.2.2.4, nulling the average error ($\alpha = 0$),

$$\Delta\Theta_m = \left(1 + \frac{\cos\Theta_i}{\cos\Theta_o}\right)\Delta\alpha \qquad (7)$$

whence, using Eq. (6),

$$\Delta\alpha = \frac{0.15\lambda/D}{1 + \dfrac{\cos\Theta_i}{\cos\Theta_o}} \qquad (8)$$

To assess typical values, let $\lambda = 0.5$ μm and $D = 7.5$ mm, whence the numerator becomes equal to 10^{-5}. When $\Theta_o = 0$, the denominator reduces to 1 $< 1 + \cos\Theta_i < 2$. When $\Theta_o \approx \Theta_i$, the denominator ≈ 2. In reflective diffraction, $\Theta_o \neq \Theta_i$, for that is pure reflection; the grating period $d = \infty$, by Eq. (4) of Section 3.2. Taking the denominator ≈ 2, then $\Delta\alpha \approx 5 \times 10^{-6} \approx 1$ arc second! Another way to look at the tolerable angular error is that over any 10 mm distance on the surface, it is to exhibit no more than $5 \times 10^{-5} = 0.05$ μm $= \frac{1}{10}$-wave departure from perfection at the selected wavelength of 0.5 μm. It need be emphasized that this was estimated on the basis of a maximum tolerable spot growth or misplacement of 20% of its FWHM. If a more stringent criterion applies, correspondingly tighter tolerances apply.

When additional media and surfaces become operative, analysis need be extended to include the corresponding refractive errors, applicable to both reflective and transmissive systems. The diffractive error was explored above for the case of a reflective hologram. A refractive error could develop from the addition of a protective transparent member having nonparallel surfaces.

Consider beam misplacement by a prism having near-parallel surfaces, a very shallow wedge angle. From Eq. (33) of Section 4.2.2.4, ray propagation through a prism with a small wedge angle $\Delta\beta$ (where $\sin\Delta\beta = \Delta\beta$) is expressed by (with subscript p for prism)

$$\sin\Theta_{ip} + \sin\Theta_{op} = n\,\Delta\beta \qquad (9)$$

where n is the refractive index of the medium immersed in $n_o = 1$ (air), Θ_{ip} is the incident ray angle with respect to incident surface normal, and Θ_{op} is the output ray angle with respect to the output surface normal. Representing a prism of small apex angle $\Delta\beta$ and input and output ray angles with respect to their surface normals, define $\Delta\Theta_{op} = \Theta_{op} - \Theta_{oi}$, where $\Delta\Theta_{op} =$ change in output angle due to the prism, and Θ_{oi} = unchanged output angle (collinear with input ray). One can then show that $\Theta_{oi} = \Delta\beta - \Theta_{ip}$, whence

$$\Delta\Theta_{op} + \Delta\beta = \Theta_{ip} + \Theta_{op} \tag{10}$$

Equation (9) relates this right-hand term to $n\,\Delta\beta$. Assuming first small input and output angles, then $\Delta\Theta_{op} + \Delta\beta = n\,\Delta\beta$, yielding

$$\Delta\Theta_{op} = \Delta\beta(n - 1) \tag{11}$$

To derive the angular error at any Θ_{ip}, we return to the general expression of Eq. (9) with modified notation, following the sign convention of Section 3.3

$$\sin\Theta_i + \sin(-\Theta_i + \Delta\beta + \epsilon) = n\,\Delta\beta \tag{12}$$

in which the subscript p for "prism" is understood. Referring to Fig. 3.2b, the nominal undeviated output is $-\Theta_i = \Theta_o$. When the substrate of Fig. 3.2b assumes a small wedge angle $\Delta\beta$ (apex on "top"), the output angle is reduced by this angle with respect to its surface normal and reduced further by deviation ϵ, which we seek to determine.

Employing series expansion for the sine of an angle and truncating beyond the cubic term, Eq. (12) becomes

$$\Theta_i - \Theta_i^3/6 - \gamma + \gamma^3/6 = n\,\Delta\beta \tag{13}$$

in which

$$\gamma = \Theta_i - (\Delta\beta + \epsilon) \tag{13a}$$

Letting

$$c = \Delta\beta + \epsilon \tag{13b}$$

where c represents the small differential components, γ^3 reduces to

$$\gamma^3 = \Theta_i^3 - 3\Theta_i^2 c + (3\Theta_i c^2 - c^3) \tag{14}$$

Ignoring the terms (in parentheses) beyond first power in c, and substituting the remaining significant terms into Eq. (13),

$$c - \frac{\Theta_i^2 c}{2} = n\,\Delta\beta \tag{15}$$

Finally, substituting for c from Eq. (13b) we obtain the approximated expression for the general error component,

$$\epsilon = \Delta\beta \frac{n - 1 + \Theta_i^2/2}{1 - \Theta_i^2/2}$$

$$= \Delta\beta \left(\frac{n}{1 - \Theta_i^2/2} - 1 \right) \qquad (16)$$

Note that when $\Theta_i \to 0$, these equations reduce to Eq. (11).

Evaluating for the case of $\Theta_i \approx 45° = \pi/4$ and $n = 1.5$, Eq. (16) reveals

$$\alpha(\pi/4) = \Delta\beta \frac{0.5 + 0.308}{1 - 0.308} \approx 1.17 \Delta\beta \qquad (17)$$

That is, the error component is about as large as the wedge angle and about double that at normal incidence. The approximations taken in this derivation may be validated when this value ($1.17 \Delta\beta$) is compared to that determined from solution of $\sin \Theta_{op}$ of Eq. (9). For $\Delta\beta = 10^{-3}$, applying procedure of Eq. (12) it yields $\epsilon(\pi/4) = 1.13 \Delta\beta$, which is lower than that determined from Eq. (16) by only 3.4%. When $\Theta_i = 30° = \pi/6$ and $n = 1.5$, Eq. (16) yields $\epsilon(\pi/6) = 0.738 \Delta\beta$, a significant reduction in error, considering that the minimum value is $0.5 \Delta\beta$ at normal incidence. Validating again with Eq. (9), the precise value for $\epsilon(\pi/6)$ is only 1.1% lower than that approximated from Eq. (16).

A further simplification of Eq. (16a) is derived by taking $1/1 - x \approx 1 + x$ for small x and normalizing the result at $\epsilon(\pi/6)$. This yields

$$\epsilon_o \approx \Delta\beta[(n - 1) + 0.58n\Theta_i^2] \qquad (18)$$

$$= \Delta\beta(n - 1) + \Delta(\Theta_i) \qquad (18a)$$

where

$$\Delta(\Theta_i) = 0.58n\Theta_i^2 \, \Delta\beta \qquad (18b)$$

Although less accurate over a wide range of Θ_i, this form lends itself better to subsequent development.

Applied to the reflective hologram, a transparent member having an angular surface error $\Delta\beta$ will divert the input ray, and after reflection, will divert it again. Effectively, $\Delta\beta$ is doubled in reflection through double-pass, yielding from Eq. (18),

$$\Delta\Theta_r = 2[\Delta\beta(n-1) + \Delta(\Theta_i)] \tag{19}$$

Using the 20% criterion on angular tolerance represented by Eq. (6) and letting $n = 1.5$,

$$\Delta\beta \approx 0.15\lambda/D - 2\Delta(\Theta_i) \tag{20}$$

is the tolerable angular deformation of an overcoating transparent member in the vicinity of D. Using the typical values selected following Eq. (8) (and taking $\Theta_i = 0$), this deformation is about twice as tolerant as that of the diffracting surface itself. That makes its acceptable deformation (at $\Theta_i = 0$) about 2 arc seconds, corresponding to approximately $\frac{1}{5}$-wave departure from perfection (at the wavelength of 0.5 μm) over any 10-mm surface subtense. A finite Θ_i could deplete the tolerance significantly, and must be determined per prior discussion.

Without a priori information on the form and orientation of the deformations, one may assume that the two are uncorrelated, yielding a total error which is the root sum square (rss) of the individual ones. If known to be in the same direction, they add, yielding a worst case from Eqs. (7) and (19) for the reflective hologram, a total angular error of

$$\Delta\Theta_{tr} = \left(1 + \frac{\cos\Theta_i}{\cos\Theta_o}\right)\Delta\alpha + 2[(n-1)\Delta\beta + \Delta(\Theta_i)] \tag{21}$$

in which $\Delta\alpha$ is the diffractive surface angular error and $\Delta\beta$ is the refractive member wedge angle, both in the same region of the illuminated aperture D. The bracketed term on the right is Eq. (18) or its more rigorous expression, Eq. (16).

5.3.2.3. Surface Deformation in Transmission

Conversion of deformation criteria from reflection to transmission is relatively straightforward, following expression of some related factors. As in reflection, there are two parameters that determine and divert ray paths:

1. Diffraction at the hologram surface
2. Refraction effects by the substrate and protective material, if used

We must first assure that the supporting and protective media are sufficiently homogeneous that their index of refraction n be considered constant. This is generally valid for relatively thin members having short optical paths. In those cases where the supporting or protective members are relatively thick, such as represented in Fig. 4.21, index homogeneity is sustained with the use of appropriate-quality optical materials.

Another condition is that the transparent medium be either thin enough or that the ray paths are sufficiently normal to their surfaces that the diffractive and refractive effects are in essentially common regions. Otherwise, the consequences of the two effects on a single spot need be determined from disparate regions. This could develop, for example, for incident or diffracted beams at $\approx 45°$ to nominally flat and parallel surfaces which are separated by a thick glass substrate.

As in reflection, we take the average error to be zero; otherwise, the problem is described better by substrate wobble. From Eq. (29) of Section 4.2.2.4 and nulling the average error, we obtain the ray deviation $\Delta\Theta_t$ due to a localized angular deviation of the grating $\Delta\alpha$:

$$\Delta\Theta_t = \left(1 - \frac{\cos\Theta_i}{\cos\Theta_o}\right)\Delta\alpha \qquad (22)$$

As expressed earlier, this equation was developed on the basis of parallel supporting surfaces, a valid criterion for substrate wobble, in which the entire substrate undergoes the angular deviation $\Delta\alpha$. However, in this situation of surface deformation, there develops a high probability toward noncongruent surface undulations which form a first-order effect of localized thin wedges which are superimposed on and about the general region of the aperture D. Thus we add the thin prism component from Eq. (18) to form a total transmitted ray deviation angle (through a transmission hologram):

$$\Delta\Theta_{tt} = \left(1 - \frac{\cos\Theta_i}{\cos\Theta_o}\right)\Delta\alpha + (n - 1)\,\Delta\beta + \Delta(\Theta_i) \qquad (23)$$

in which $\Delta\alpha$ is the component of angular error where the supporting surfaces may be assumed parallel, and $\Delta\beta$ [also contained in $\Delta(\Theta_i)$] is the small wedge component of angular surface error. As for Eq. (21), this represents the maximum beam deviation due to spatially correlated tilt and wedge components. Random combinations merit an rss assumption, following our earlier discussion.

Comparing Eq. (23) for transmission holograms and Eq. (21) for reflection holograms, we see that although similar, significant distinctions exist. As in wobble error, only in the transmission mode can the plane-parallel angular error $\Delta\alpha$ be nulled with $\Theta_i \approx \Theta_o$. Since most surface errors of subtense in the vicinity of aperture size D are pseudorandom and cannot be considered as fortuitous plane-parallel deviations, the wedge component probably dominates. And this factor is one-half as severe in transmission, due to its single-pass versus the double-pass deviation of reflection.

For $n = 1.5$ and $\Theta_i = 0$, then, $(n - 1) \, \Delta\beta = 0.5 \, \Delta\beta$ is the residual output beam displacement angle due to wedge alone. Comparing this to the maximum tolerable deviation of Eq. (6) *for a 20% spot deviation criterion which can be considered moderate* [see the discussion that yielded Eq. (6)], then

$$\Delta\beta = 0.3 \, \frac{\lambda}{D} \qquad (24)$$

which is twice that for reflection, represented by Eq. (20) at $\Theta_i = 0$. At $\lambda = 0.5 \, \mu$m and $D = 7.5$ mm, $\Delta\beta = 20 \times 10^{-6} \approx 4$ arc seconds. This corresponds to 0.4-wave departure from perfection (differential angle between the two surfaces) over any 10-mm surface subtense at the wavelength of 0.5 μm. Again, a finite Θ_i could deplete the tolerance significantly, and must be determined following our earlier discussion.

A method of compensating for surface deformation or substrate inhomogeneity proposed by Bjorklund and Sincerbox (Bjo) follows a procedure employed for correcting conventional optical components (Smi). It entails recording the hologram with the conjugate to the aberrating one. Techniques are discussed for achieving conjugation and overcoming complications, especially by dynamic methods that yield conjugate waves in real time.

5.4. WAVELENGTH-SHIFT ERRORS AND THEIR CORRECTION

5.4.1. Introduction

The earlier sections in this chapter discuss primarily limitations on mechanical and structural perfection. In this section we concentrate on a uniquely optical anomaly—hologram construction at one wavelength and reconstruction at another. It is well recognized (Mei 1) that an interferometrically generated hologram (having undergone no topological shift) will reconstruct a stigmatic image when reilluminated with the precise wavefront (or its conjugate) used for construction—independent of the contour of the substrate. Complications arise from two general sources: (1) a shift in mechanical characteristics (distances, angles, surface quality) between exposure and reconstruction, and (2) use of a reconstruction wavelength different from that used in construction. Surely, these two can coexist. In fact, a powerful general corrective approach, subsequently discussed, is to adjust the parameters of (1) to complement (in part) the consequences of (2).

Although this wavelength shift could result from instability of light-source oscillation (such as from a lasing diode subsequently discussed), it is often imposed far more profoundly with a significant disparity between recording and reconstruction wavelengths. This results typically from incompatibility of holographic recording media to the radiation at the desired final wavelength, that is, the need to record at the wavelength of high media sensitivity (often <450 nm) and to reconstruct at the wavelength of economical and appropriate radiant source availability (often >700 nm). Section 4.2.2.2.1 describes a unique corrective procedure utilizing the properties of the Rowland circle applied to the concave Holofacet scanner. It is not, however, a general solution.

It is quite clear (deferring confirmation until later) that the problem (also appearing in Section 4.2.2.6.4) merits careful attention. A motivation for laser-based scanning and recording is the realizable high integrity of its focal spots, forming the cumulative attribute called high resolution. Thus total wavefront error exceeding $\lambda/2$ departure from perfection is seldom justified or accepted. As an index of resolution perturbation, it may be appreciated that a "standard, high-quality" $\lambda/4$-wave error can enlarge a typical spot at FWHM by approximately 20%, and that a $\lambda/2$-wave error can more than double the size of an original diffraction-limited spot. It will be seen that under typical conditions, uncompensated wavelength-shift errors can far exceed these perturbations.

While the general wavelength-shift problem does encompass a vast scope of spatial parameters and conditions, some characteristics of holographic scanning tend to reduce the variables significantly, creating a potential for practical control. First, we are concerned most often with single conjugate-point imaging. That is, the holographic optical element (HOE) mounted on an articulated substrate serves to image *only* a single point source to a single point image. Note that a collimated input or output is a single point source or image at infinity. It will be seen that almost all analyses of wavelength-shift errors are, fortunately, limited to this single-point imaging problem, recognized as the fundamental building block of an otherwise more complex (superimposed point) image space. This more complex image space is generally not encountered in holographic scanning. Second, of the HOEs applied to scanning, those exhibiting no optical power on flat substrates (plane linear diffraction gratings) experience no aberration due to wavelength shift, only a change in diffraction angle. Third, of those exhibiting lenticular power where radial symmetry is sustained, the aberration is stationary with scanning, allowing uniform (at least partial) compensation with a fixed correction process. Finally, it will be seen that certain corrective procedures are more effective for off-axis holograms; which are far more prevalent in holographic scanning than on-axis ones. As represented in Sections 4.3 and 4.5, the in-line zone lens (on-axis hologram) serves primarily for the expression of diffractive scanning fundamentals. To clarify this subject, an awareness of the basics of classical aberration theory is useful for an appreciation of subsequent analyses and consideration.

5.4.2. Hologram Aberration Theory

Most comprehensive analyses of optical aberration (holographic wavelength-shift error included) is based on Seidel aberration theory. The precision of determination of image spot distribution from ray trace formulas, which are dominated by the sines of field angles, depends on the degree of approximation of these sine functions. To allow closed-form expression of aberrations to varying degrees of accuracy, the series expansion of $\sin \Theta = \Theta - \Theta^3/3! + \Theta^5/5! - \Theta^7/7! + \cdots$ is truncated at the desired degree of accuracy, considering the resulting complexity. The paraxial assumption is that $\sin \Theta = \Theta$, known as first-order theory. Truncating after the third-order term is third-order theory, and so on. The most prevalent compromise between the inadequacy of first-order theory and the complexity of higher-order theories is the third-order approximation. To appreciate its practical utility, consider the angle $30°$, for which $\sin \Theta = 0.50$, $\Theta = \pi/6 = 0.5236$, $\Theta^3/3! = 0.0239$, and $\Theta^5/5! = 0.00033$. The fifth-order term is a minor correction to the fourth decimal place.

An imperfect spherical-like wavefront brought to an image "point" (which may be diffracted from a hologram) can be represented by $e^{-ik(r+\Delta)}$, where r is the distance from the image point to the hologram and Δ is the deviation from perfection of a spherical wavefront. In an x-y-z rectangular coordinate system (hologram in x-y plane), when the off-axis object or image point is confined to a single $(x$-$z)$ plane (note adaptation to rotational symmetry), the deviation from a perfect wavefront may be represented as (Cha 1, Cha 2, Meh)

$$\Delta = -\tfrac{1}{8}(x^2 + y^2)^2 S + \tfrac{1}{2}x(x^2 + y^2)C - \tfrac{1}{2}x^2 A \qquad (1)$$

$$= \Delta_S + \Delta_C + \Delta_A \qquad (1a)$$

in which S, C, and A are equivalent to the Seidel coefficients for spherical aberration, coma, and astigmatism, respectively, and the corresponding Δs represent the individual deviations. Although Seidel coefficients exist for additional errors known as curvature of field and distortion, these two are not operative in scanned single-point image systems. Denoting $\rho = (x^2 + y^2)^{1/2}$ in the hologram plane oriented at angle Φ to the axis, then $x = \rho \sin \Phi$ and $y = \rho \cos \Phi$, whereupon we see that spherical aberration is proportional to the fourth power of ρ, coma to the third power, and astigmatism to the second power of ρ. As employed here, ρ is equivalent to half the illuminated aperture, $D/2$.

The hologram is made with object and reference beams oriented at angles α_o and α_r with respect to hologram normal, and utilized with a reconstruction beam at angle α_c forming a diffracted image beam at angle α_i with respect to surface normal. Subscripts o, r, c, and i are those frequently used to describe the four hologram beams. In particular, note the distinction between these subscripts i and o (image and object) and those used elsewhere in this volume for input and output beams. All source and image points are oriented at corresponding distances R_o, R_r, R_c, and R_i from the hologram. Using this nomenclature and the ratio of reconstruction wavelength to reference wavelength

$$\mu = \lambda_c / \lambda_r \tag{2}$$

the aberration coefficients S, C, and A are expressed as (Cha 2)

$$S = \frac{1}{R_c^3} \pm \mu \left(\frac{1}{R_o^3} - \frac{1}{R_r^3} \right) - \frac{1}{R_i^3} \tag{3}$$

$$C = \frac{\sin \alpha_c}{R_c^2} \pm \mu \left(\frac{\sin \alpha_o}{R_o^2} - \frac{\sin \alpha_r}{R_r^2} \right) - \frac{\sin \alpha_i}{R_i^2} \tag{4}$$

and

$$A = \frac{\sin^2 \alpha_c}{R_c} \pm \mu \left(\frac{\sin^2 \alpha_o}{R_o} - \frac{\sin^2 \alpha_r}{R_r} \right) - \frac{\sin^2 \alpha_i}{R_i} \tag{5}$$

in which the upper ($+$) sign denotes the virtual image and the lower ($-$) sign, the real image. Their positions satisfy the equivalent of the thin lens (gaussian) equation $1/R_i = 1/R_c \pm 1/f$,

$$\frac{1}{R_i} = \frac{1}{R_c} \pm \mu \left(\frac{1}{R_o} - \frac{1}{R_r} \right) \tag{6}$$

The gaussian relationship between their angles (Cha 1) is also derivable from the grating equation as

$$\sin \alpha_i - \sin \alpha_c - \pm \mu (\sin \alpha_o - \sin \alpha_r) \tag{7}$$

accounting for sign-convention differences in the referenced works. For the sign convention adopted here, the transmission grating (real image) equation is expressed as $\sin \alpha_i + \sin \alpha_c = \mu (\sin \alpha_o + \sin \alpha_r)$. The image and reconstruction beam angle sines increase in proportion to the wavelength ratio μ. The angular changes experienced for a disc scanner (Kra 7) and for one similar to Fig. 4.22 (Kuo) were evaluated in this manner.

5.4.3. Abberration Evaluation and Balancing Method

In a comprehensive work by Latta (Lat) appears a representation of the seriousness of lenticular hologram aberration and an approach to its adjustment. Utilizing computerized analyses based on the foregoing equations, both in-line and symmetric off-axis holograms were evaluated over a wide range of conditions, including those adaptable to scanning. For example, image points were selected as real, having f numbers ranging from $f/1$ through $f/50$, $\lambda_r = 633$ nm and Δ_c to ± 300 nm. Some germane results are summarized as follows: For $10°$ symmetric off-axis image and reconstruction beam angles, and using collimated reference/reconstruction beams, an $f/20$ focal spot imaged 200 mm from the hologram will suffer $\lambda/2$ total aberration when $\Delta\lambda_c \approx \pm 20$ nm. When $\Delta\lambda_c = \pm 100$ nm, the total wavefront error is about 2λ. An $f/10$ spot will aberrate $\lambda/2$ when $\Delta\lambda_c \approx \pm 7$ nm, and when $\Delta\lambda_c = \pm 100$ nm, the total wavefront error is about 10λ. When the off-axis illumination is extended to $25°$ symmetric, maintaining all else the same, the $f/20$ focal spot will suffer $\lambda/2$ aberration when $\Delta\lambda_c \approx \pm 7$ nm, and when $\Delta\lambda_c = \pm 100$ nm, the total wavefront error is about 10λ. An $f/50$ focal spot will suffer $\lambda/2$ aberration for $\Delta\lambda_c \approx \pm 25$ nm, and about 2λ error at ± 100-nm wavelength deviation. Although not explicitly evaluated, extrapolation of the $25°$ offset data indicates that an $f/100$ focal spot will experience $\lambda/2$ aberration for as little as ± 100 nm shift in reconstruction wavelength (from 633 nm). The anticipated significance of this problem is clearly represented in these data.

The method for aberration balancing invokes the following conditions upon the aberration components of Eqs. (1), (3), (4), and (5):

$$\left| \Delta_S \right|_{max} = -\left| \Delta_A \right|_{max} \tag{8}$$

$$\Delta_C = 0 \tag{9}$$

A complementary contribution of Δ_A with Δ_S requires operation off-axis; for on-axis, only spherical aberration remains. Thus, interestingly, the more off-axis the operation, the more effective the balancing, making the null in aberration practically constant for reconstruction beam angles in the range $10° < \alpha_c < 35°$. Uncompensated, the aberration increases rapidly with α_c. In one case, doubling the angle (from $10°$ to $20°$) trebles the aberration, and tripling the angle (from $10°$ to $30°$) multiples aberration by a factor of 7.

To allow a tractable analysis, points are taken off-axis in the x coordinate only. Balancing is instituted with two of the four hologram beams specified independently and the remaining two determined therefrom. A practical choice (adaptable to scanning) was made: specifying the real image beam and the reconstruction beam, and determining the required object and reference beams.

Utilizing this constancy of image and reconstruction beams, several consolidating substitutions into Eqs. (3), (4), and (5) ease their rearrangement to satisfy balancing equations (8) and (9) and gaussian image equations (6) and (7). These four new equations, "after considerable algebraic tedium," yield the object beam angle α_o and a polynomial in R_o in terms of reconstruction and image beam parameters.

Some results are expressed in computed examples of wavelength shift of $\mu \approx 1.3$, from $\lambda_r = 488$ nm to $\lambda_c = 633$ nm. Offsets of (real) image α_i and reconstruction beam α_c are maintained symmetric, computed through angles from 5 to 35°. Over a range of $10° < \lambda_c < 35°$, an $f/5$ output beam imaging to a point $R_i = 100$ mm from the hologram experiences total aberration of about 0.2λ. When the f-number is increased to $f/10$ (hologram aperture D halved), the aberration is reduced to $<0.012\lambda$. Extrapolating to higher f numbers, the aberration under the same conditions becomes vanishingly small. The reason for incomplete cancellation per Eq. (8) is that, as expressed earlier, spherical aberration varies as D^4, whereas astigmatism varies as D^2. Thus their aberration functions fail to be complementary over the full aperture.

5.4.3.1. Balancing for Nonsymmetric Input and Output Beam Angles

All evaluations thus far were made for symmetric reconstruction and image beam angles about the hologram normal. Further computation was conducted with departure from symmetry while maintaining a constant 10° differential between the reconstruction beam direction and the direction of the diffracted image point. Except for this variation, conditions were the same as those expressed above, utilizing an $f/5$ output cone. Observations were significant. First, aberration is reduced drastically when reconstruction beam $\alpha_c \approx 0$, that is, normal to the hologram. When collimated, this represents radial symmetry for a disk scanner; when the reconstruction beam is diverging, it represents radial symmetry for the configuration of Fig. 4.22. Second, aberration approaches a peak for the condition of symmetric illumination/reconstruction. The actual maximum ($\approx 4x$ that when symmetric) occurs when the image beam exits close to normal and the reconstruction beam is at $\alpha_c \approx 10°$. This operation is similar to that represented in Fig. 4.23, with a shallower angle of the input beam.

5.4.3.2. Two-Step Hologram Construction

An aberration nulling procedure developed by Lin and Doherty (Lin, Sin) may be applied to holographic scanning under wavelength-shift conditions. It is a two-step process in which a first hologram H_1 is recorded at the desired final (usually longer) wavelength λ_1 on an appropriately sensitive material (such as silver halide). This now serves as a master in a re-holographing process to form hologram H_2 at a different (usually shorter) wavelength λ_2 in a material sensitive to λ_2. Here the diffracted wave from H_1 which is aberrated (due to wavelength shift) at λ_2 joins the transmitted zero-order wave (now acting as reference) at λ_2 to interfere on H_2, the hologram intended for scanning. When H_2 is reilluminated at λ_1, the desired operating wavelength, the aberration is complemented and the diffracted output is stigmatic.

The holographic materials require control and selection for correct thickness, neither too thin nor too thick (for diffraction). If either is excessively thin, it will diffract both the desired wave and its conjugate. Both will be recorded in H_2, adding a conjugate background upon reconstruction and depleting efficiency from the desired component. If either is too thick, efficiency is impaired by departure from Bragg performance during necessary angular shifting between λ_1 and λ_2. Evaluations were conducted to estimate limits upon these factors (Lin). Also, because of the angular shifting experienced, careful control of the input and output angles is required to satisfy final angular constraints.

5.4.4. Small Wavelength Shift, as from Laser Diode

Effects could arise when there is not only a large shift between recording and reconstructing wavelengths, but when the reconstructing source generates its own small change in wavelength (Sin). A laser diode can exhibit variations of a few nanometers about nominal, primarily as a function of drive current and temperature variations. This was introduced at the end of Section 4.2.2.6.4. Such a shift can, in turn, cause two effects: cross-scan error and line-length variations. The cross-scan misplacement is a clear consequence of the wavelength shift expressed above and represented in the grating equation. This can be nulled and will be discussed subsequently. The line-length variation is, however, a subtle effect of the cross-scan tilt as resolved in the along-scan direction for nonradially symmetric systems. That is, if the scan magnification varies for different Φ as a function of wavelength, the scan magnitude will vary accordingly. In radially symmetric systems, it is uniformly unity. In the nonradially symmetric systems discussed in Section 5.2.2.3 for plane linear gratings, some of the cross-scan error is transferred to line-length change as the system departs from center scan position.

Having derived Eq. (11) in Section 5.2.2.3 as a general expression for scan linearity ratio, this characteristic is implicit there. However, Eq. (11) assumes a constant Θ_o (output angle) which results from a constant $k = \lambda/d$ in the grating equation. But k now varies with wavelength, imparting a variation to Θ_o. Performing the same trigonometric substitutions for $\cos \Theta_o$ as formed Eq. (10) from Eq. (9) in Section 5.2.2.3, an equation results which is now only a function of the input angle Θ_i, k, and the mechanical rotation angle Φ:

$$r_x = \left[\frac{\cos^2\Theta_i + 2k \sin \Theta_i - k^2}{\cos^2\Theta_i + 2k \sin \Theta_i \cos \Phi - k^2} \right]^{1/2} \cos \Phi \qquad (10)$$

By applying known wavelength-shift factors to k, the linearity ratio can be determined for any Θ_i as a function of Φ and then compared to r_x for a nominal k. For the condition of $\Theta_i = 45°$ and scan angle $\Phi = 25°$, a variation of 1 and 2 nm above 800 nm wavelength reveals 0.27% and 0.46% lengthening, respectively. Such data must be considered for each application, depending on, for example, the rate of change of wavelength. It is noteworthy that when $\Theta_i = 0$, the error is nulled.

A method was developed by D. B. Kay for nulling the cross-scan tilt portion of the problem (Kay). This is done by interposing a fixed grating between the source and the holographic scanner such that its tilt error complements the error of the scanner—at least in the central region of scan. Addressing the disk configuration illustrated in Fig. 4.33 having collimated input and output beams, another holographic grating, made identically to one of the facets, is added within the input beam path. The diffracted output of this first grating now becomes the input to the holographic scanner. With their complementary orientations, the angular change at the output is now converted to a minor translation of the beam, forming no shift of the subsequently focused image point—again, in the central portion of scan, where nulling is theoretically perfect. Kay shows data for one implementation where the margin of a 4.5-in. scan length exhibiting an uncompensated cross-scan error of approximately 50 mils/nm is reduced to approx. 0.63 mil/nm when compensated. Other precautions regarding radiation stability and uniformity of laser diodes applied to holographic scanning are expressed in this work.

Appendix 1

Historical Review of Holographic Scanning

A1.1. INTRODUCTION

It is inevitable that the extensive research which is conducted to provide continuity and rigor to a work such as this must reveal some clarification of the formative development of its technology. Thus we accede to the responsibility of that new awareness and offer here a historical review of the evolution of holographic scanning. Although such a chronological perspective may elicit a cause-and-effect progression, in reality it appears that much of the early work in independent laboratories was, in fact, independent. This study also serves to reveal some of the constraints and latitudes that shaped the patent literature. This fascinating innovative progression can surely sustain much more intensive dedication than may be accorded here. But we do provide a consolidation to a level that is otherwise uncharted.

A1.2. FIRST SIGNS

How far back do we go? In the beginning there was optical scanning and holography. By the time the first signs of holographic scanning surfaced, both disciplines had matured substantially. Surely, they were destined to be combined by those active in both fields. And they were, after 1964. The more notable it is, therefore, that for all practical purposes, holographic scanning was invented much earlier, by a person who could not have been exposed to holography—for it did not exist then.

Hollis S. Baird was acknowledged as a principal innovator in the pioneering days of television—the early 1930s. He was apparently also well oriented in the characteristics of diffractive optics, for in 1931, he disclosed use of a set of zone plates disposed on a cylindrical spinner to form the elements of the first diffractive optics scanner (Bai). Figure A1.1 is a reproduction of the cover page

171

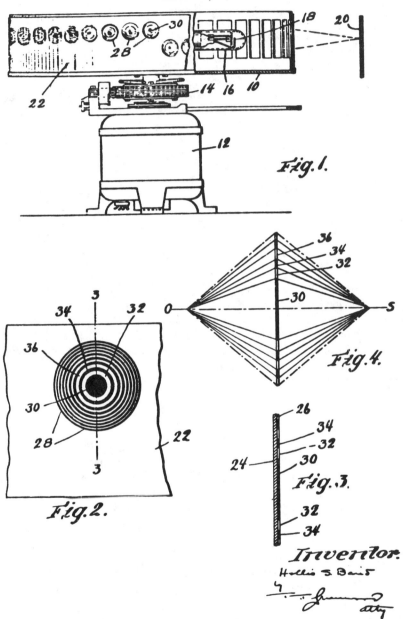

Figure A1.1. First page of U.S. Patent 1,964,474 by Baird, awarded June 14, 1934, showing zone lenses set into periphery of cylindrical drum illuminated with a "point source" and rotated to scan a raster.

of U.S. Patent 1,962,474, issued June 12, 1934 to H. S. Baird. Not only did he generate a two-dimensional raster, but he anticipated the use of the thin zone plates embossed on a thin flexible band which forms the cylinder, and he allowed for photographic reproduction of the zone lenses from a larger master zone plate, for uniformity and economy. Further, he anticipated changing focal length with a change in image reduction ratio from the master zone plate.

So dominant was that revelation of prior art that it has, in fact, frustrated subsequent reinvention after holography became known in its present form. This is reported here in several instances. The first is known because of the relationship of the present author to CBS Laboratories in Stamford, Connecticut, during these formative years (serving as Staff Scientist to CBS Labs, Division of CBS, while Dennis Gabor was Staff Scientist to all of CBS). A driving force behind holographic scanning research was the then-Vice-President of the Laboratories, Renville H. McMann. A pioneer in television development in his own right, on November 24, 1967, he filed for patent coverage (McM 4) of configurations very similar to Baird's cylindrical scanner and Jenkins' earlier (Jen 1) disk TV scanner which filled Nipkow's famous (ca. 1884) spiral holes in a disk with lenses, as in Fig. A1.2. This prior art frustrated issuance of that patent. The time line is indelibly impressed on Ren McMann, for on August 1966, his eyes were injured through accidental laser exposure during that experiment.

A major factor leading toward holographic scanning at CBS Laboratories was the earlier dedication to laser scanning in general, for the purpose of advancing reconnaissance phototransmission. Aerial photographs having extremely high resolution and image quality are scanned, digitized, and transmitted to a central station for reconstruction and interpretation. In the 1964–1965 era, intensive research was initiated toward leapfrogging the best performance of the then-dominant cathode-ray-tube (CRT) systems (notably represented by the rotating phosphor anode drum "line scan tube" of CBS Laboratories).

On September 17, 1964, this author responded to a request for proposal from the Air Force for an "Optical Spot Size Study for Data Extraction from a Transparency." Figure A1.3 is a reproduction of the scanner illustration—a lens-disk assembly composed of an array of equal (angularly) spaced positive lenses mounted about the periphery of a disk which is rotated about its axis. A laser beam is expanded and normalized (to present a uniform flux distribution). Upon disk rotation, the lens apertures traverse the collimated (radially symmetric) flux field to generate an arced locus of a (2.5-μm) focal point on the flat image plane. The program was awarded and the research was conducted rapidly and in 1965 reported broadly (Bei 13, Izz). It may be recognized as a precursor to much early work in holographic scanning, exemplified in Fig. A1.2 as Fig. 1, and similar to Fig. A1.5. Ren McMann and I considered several times the trade-offs of replacing the lenses with reflective paraboloids or ellipsoids, and with

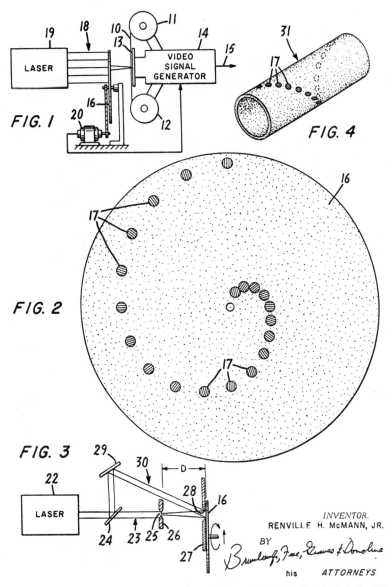

Figure A1.2. Illustration from patent filing November 44, 1967 by R. H. McMann for "Optical scanner," showing "holographic image of pinholes" oriented about periphery of a cylinder and in a spiral upon a disk, to generate a raster upon reillumination and rotation.

174

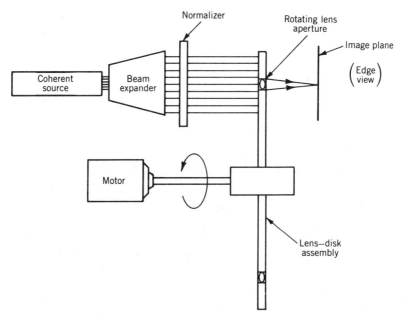

Figure A1.3. Illustration from "Optical Spot Size Study for Data Extraction from a Transparency," prepared by L. Beiser, September 17, 1964.

transmissive or reflective holograms. This program initiated a more recognizable utility in that the phrase "beam expander" appearing in Fig. A1.3 was originated by this author in 1964. This is the first known use of that descriptor, which now almost completely supersedes the classic phrase "inverted telescope." With many program variations and major developmental by-products for advanced phototransmission and wideband recording systems, this early work in laser beam scanning by CBS Laboratories became widely distributed to the image and data handling communities (Bei 6) and to the public. The name of the most celebrated system, Compass Link, was used by President Johnson in a televised conference on Southeast Asia, and in 1970, Compass Link was described in a news release (Gou) regarding its deployment in the Middle East. This backdrop of activity at CBS Laboratories is the motivation for research there in holographic scanning, directed toward advancing the extremely high performance in resolution and speed provided by their reconnaissance-type rotating polygon systems.

This brings us to Dennis Gabor. Certainly, one would not expect this masterful innovator to sit still during this exciting era of electro-optical discovery and progress, especially with his deeply perceptive orientation in image information handling. While much of our documentation is in the form of private communications, it does provide a backdrop for the directions that developed as inevitable consequences of adaptation of diffraction optics to scanning. Our first record of such work appears in an early 1965 memo to Peter Goldmark, then President of CBS Laboratories. Figure A1.4 is a reproduction of the first of three pages of "Report No. 9—A New Method of Optical Scanning" (Gab 1),

TO: DR. P.C. GOLDMARK

FROM: D. GABOR   STAMFORD, April 12, 1965

REPORT NO. 9.   A NEW METHOD OF OPTICAL SCANNING.

Like the device described in my first report, this is an application of the new imaging methods which have become possible by the availability of coherent light with high intensity.

The basic idea is producing a zone lens which runs in an ultrasonic sound trough with the velocity of sound. In almost all liquids this is 900 - 1500 meters/sec. A zone lens is produced by producing a sufficiently sharp pulse with a piezocrystal, with a frequency which varies with time according to a saw-tooth curve, i.e. linearly with time. Each pulse produces a running scattering centre, which produces a cylindical wave, and interferes with those emitted by the other scattering centers. In the sketch the spaces between the scattering centres have been shown black. In reality these transmit the wave in its original direction, and this "zero order" wave will have to be cut out by methods to be explained later.

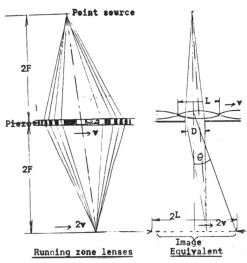

A zone lens is always at the same time a positive and a negative lens. It will be best to make its distance from the point source (laser focus,) twice its focal length, so that one of the images always coincides with the source and gives no further trouble.

As the velocity of sound in liquids and solids is very large, the zone lenses are very long. In water a millisecond corresponds to 150 cm, and a TV scan line period, 1/15,000 sec to 10 cm. I propose therefore to put a smaller window of length D before it; this is the actual length of the trough. The lens period itself is L.

The basic relations. The image, at 2F from the lens has a scannable length of 2L. The maximum deflection angle is

$$\theta = L/2F = \lambda/d \qquad\qquad 1.$$

where d is the distance between fringes. (Scattering centers.) The number of fringes simultaneously before the window D is therefore

Figure A1.4. Reproduction of page 1 of three-page memo by D. Gabor on April 14, 1965, showing acoustooptic traveling-lens form of diffraction optics scanner.

in which Dennis (everybody at the Labs called him that) disclosed a "traveling lens" (Fos) or "chirp" (Bed) scanner, an acoustooptic technique which, as far as we know, did not surface in the literature before 1970 (Fos). Although not a holographic scanner of current interpretation, it does provide controlled motion of a stationary diffraction grating—stationary in the sense that the grating structure remains fixed with respect to its own coordinates while it traverses a linear path very rapidly.

A real holographic scanner was subsequently described in a memo from Dennis to Peter Goldmark, dated April 3, 1967 (Gab 2), reproduced here as Fig. A1.5. Of relatively poor quality, it is the best of two copies in our files. Having communicated with the CBS patent office at the Tech Center (formerly CBS Laboratories) in Stamford, Connecticut, this may be the best now available. Apparently, he and Peter Goldmark had been talking privately about "this project," for which Dennis considered this realization an optimal one. It is a transmissive disk scanner, radially symmetric (see Section 4.1.2) and operating much like an array of positive lenses equally spaced about the periphery of a disk and rotated about its axis, per Fig. A1.3. It is also overilluminated (for full aperture utilization and zero retrace time), although not limited to that operation (see Section 2.8.1). Gabor's practical concerns appear in his equal object-image distances to minimize defocus due to disk flutter and attention to hologram efficiency, expressing the need for "deep" holograms (now called "thick") to attain high efficiency. Although there is evidence of some disclosure for patent application, it appears not to have gone past the talking stage. It was, however, followed by experiment. On September 1, 1967, a memo to Ren McMann reported successful test of that configuration using a bleached silver halide plate as the hologram. The illustration accompanying that memo is reproduced here as Fig. A1.6. The reconstructed spot was expressed as a diffraction-limited $f/4$, having a 6-μm diameter to the null of the first ring (approximately 3 μm to FWHM). Also, the position of the reference beam source must be held quite accurately (for this high resolution), determining that a 4-mil displacement begins to show deterioration. With accurate positioning, the scanned spot (in this radially symmetric system) was reported to maintain size constant throughout the full rotation of the disk (on a 5-in.-diameter circle). Most of the above material is heretofore unpublished private communications.

A1.3. EARLY PUBLISHED SYSTEMS

The most significant public expression of work in holographic scanning in that time frame was that of I. Cindrich (Cin), then a member of the Radar Optics Laboratory of the University of Michigan—recognized where holography was revitalized by Leith and Upatnieks to world-reverberation proportions (Gab 3).

I have corrected an error in the lower figure. I had
forgotten that the "axis" is a broken line.
D. Gabor, April 8, 1967

TO: DR. P.C. GOLDMARK

FROM: D. GABOR

April 3, 1967

REPORT NO. 3. HOLOGRAPHIC SCANNING DISC.

This is to suggest what in my opinion is the optimum realisation of this project. It has two characteristic features:-

1. The source point S in the taking which is also the illumination point in operation is on the axis of rotation. Consequently the image of S is formed under identical conditions, whatever the position of the hologram-lens.

2. The distance of S from the disc is equal to the distance of the disc from the image. This makes the focusing independent of a small flutter of the disc in axial direction.

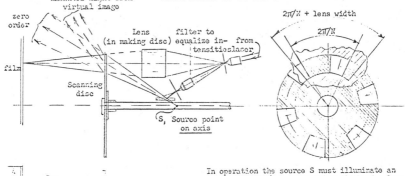

zero order

unwanted light from virtual image

film

Lens   filter to
(in making disc) equalize in- from
tensities laser

$2\pi/N$ + lens width

$2\pi/N$

Scanning disc

S, Source point on axis

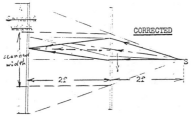

CORRECTED

scanned width

2f

2f

S

In operation the source S must illuminate an arc equal to $2\pi/N$ + the angular width of a lens. By this each lens is fully illuminated during the whole time of its scan, which is $1/N$ of a rotation period. I have drawn the sketch for N = 6, angular width of lenses 20°. This gives an illumination angle of 60 + 20 = 80°, which is rather inconveniently large. I recommend N = 12. illuminating angle 40°. ~~Even this gives a wide scan, because the scanning width in this arrangement is twice the arc described by a lenticule.~~ With a film width of 65 mm this gives an arc of ~~34.5~~ mm, a periphery of 12 x ~~65.9~~ = ~~390~~ mm and a diameter of ~~62.5 mm~~ or ~~2~~ inch. 125 mm / 5

In the making of the hologram it is important to equalize the intensities in the beams coming from the point S and through the lens. This can be done with a polaroid. The lens must be of high quality.

The simple theory gives a maximum efficiency of 6.25% for a black-and-white transmission hologram, but this is not quite right if a 649F plate is used, which gives a deep hologram. Here the efficiency can be very much higher, (theoretically 100%) and it is reported that efficiencies of the order 60% have been realised.

D. Gabor

Figure A1.5. Reproduction of one-page memo by D. Gabor on April 3, 1967 (addendum April 8) describing a ''holographic scanning disk.''

178

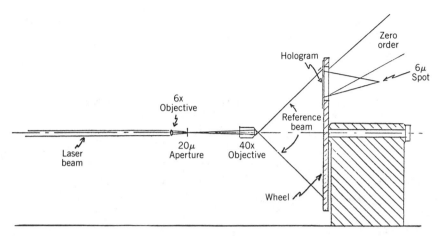

Figure A1.6. Reproduction of illustration of experiment reported September 1, 1967 describing successful test of Gabor's earlier configuration. "6 μm spot," is diameter to first null, corresponding to approximately 3 μm FWHM, which was sustained throughout full scan.

In the field of holographic scanning, Cindrich documented no less dramatic a representation of this new technology, rendering masterful expression of theory and potential practice. He identified and quantified radial symmetry, analyzed the consequences of reillumination misalignment, and anticipated purposeful misalignment of the reconstruction wave to generate noncircular, though aberrated scan functions from a disk-format holographic scanner. This work, to our knowledge, merits the distinction of being the first public exposition of holographic scanning. It was received by *Applied Optics* on February 27, 1967, surely following unpublished research, and was published September 1967.

A novel disk configuration appeared in February 1969 (received by *Applied Optics* July 15, 1968) by D. H. McMahon and his associates at Sperry Rand Research Center (McM 1). Intended for use similar to that (unpublished) by Ren McMann described earlier, they generated (with 57 holograms around the rim of a disk) a 57-line raster (vertical array of horizontally scanned slightly arcuate lines) with each rotation of the disk. It was tested in a real-time TV-type arrangement in which a He-Ne laser beam was directed to scan an object to derive a signal that modulated an argon-ion laser beam which was scanned by the same holographic deflector to form a blue-green image of the object. In this work, McMahon and his associates provided analyses of resolution, considered hologram efficiency, anticipated varying the focal length of the holograms sequentially to provide depth information, and considered providing a three-color display by recording three concentric rings of holograms, each dedicated to its own optimal spectral characteristic, with focal points overlapping in the image field. The patent on this work was filed September 25, 1968 and was issued November 9, 1971 (McM 2). In addition to the above, it describes the use of tandem holographic scanners to provide x-y and x-y-z scan. Interestingly, only the tandem systems survived in the three claims, and it is noteworthy that Jenkins (Jen 1) and Baird (Bai) are both cited references—sounds like the same frustration as earlier for McMann.

In the meantime, back at CBS Laboratories, after interim dedication to creating systems such as Compass Link described earlier, work on holographic scanning was accelerated in 1968 with new awareness of increased optical efficiency in reflection with dichromated gelatin gratings and with blazed reflective holographic gratings (Dal). Thus the holograms may be deposited upon stable solid substrates such as beryllium and, as expressed in a memo by this author, then "spin the daylights out of it!" With this was initiated effort that culminated in the Holofacet systems. Early in 1969, George Stroke, then a consultant to CBS Laboratories, brought to our attention significant work by Jobin-Yvon in France, whereby they had developed stigmatic imaging from a reflective concave grating at a wavelength other than that used for exposure (Fla 2, Str 2), described in Section 4.2.2.2.1. Anticipating application of this as well as their high-quality grating expertise to holographic scanning, CBS Laboratories contracted Jobin-Yvon to fabricate holograms in accordance with this author's scanner designs. On April 29, 1969, CBS Dwg. No. B-16996 (detailing a three-faceted 3-in.-diameter reflecting holographic disk) was forwarded by Leo Beiser to Jobin-Yvon for exposure at 488 nm and reconstruction at 633 nm. It was received August 27, tested and reported September 4 as providing static optical quality (after wavelength shift) comparable to that from the best available lens apodized to the same aperture shape and f number. Dynamic tests were conducted and reported October 10 as demonstrating (for this radially symmetric system) uniformly high performance throughout the scan cycle. This predecessor to many subsequent holographic scanners is now on loan by this author to Boston's Museum of Science. The patent for a similar concave reflecting scanner appeared in 1973 (Pie 1), substantially beyond its inception in early 1969. Its U.S. filing was on July 9, 1971, long after its development at CBS Laboratories. The corresponding French application was filed almost one year earlier, during the active contract period with CBS. Characteristics of this scanner are covered in Section 4.2.2.2.1.

Soon following the introduction of the spherical concave scanner, on May 17 and 18 of 1969, this author transformed it to a convex configuration (Bei 14) so that the scanner may be formed as a relatively small sphere, in a manner similar to the high-performance pyramidal polygon employed on Compass Link, with the output beam directed normal to the rotating axis. Pages 1 and 4 of the disclosure notes (read and understood by George W. Stroke) are reproduced here as Fig. A1.7. That became the first patented holographic scanner (Bei 1), issued October 19, 1971. Its fabrication and test are described in Section 4.2.1.1.

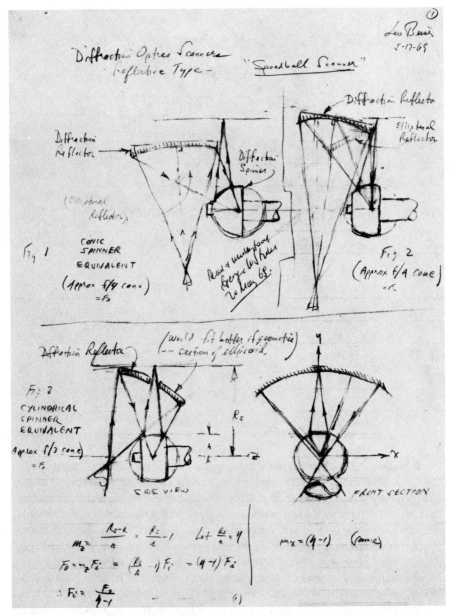

Figure A1.7. Reproduction of pages 1 and 4 of notes by L. Beiser describing the convex holographic scanner which became known as the Holofacet system. (*a*) May 17, 1969, different illumination methods and section utilization of spherical substrate. (*b*) May 18, 1969, Fig. 4, pictorial of scanner with ellipsoidal reflector and reflective spherical scanner disposed in cupped film transport system.

181

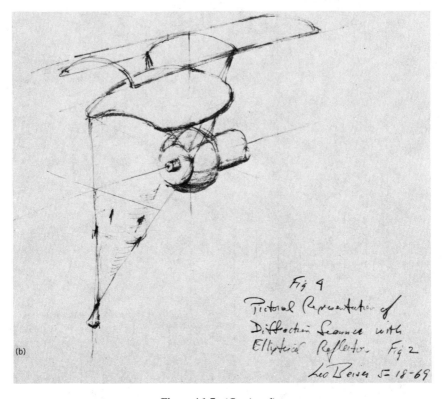

(b)

Figure A1.7. (*Continued*)

Development followed rapidly under partial Air Force support during 1971–1972. A 4-in.-diameter beryllium sphere having 12 Holofacets disposed about its periphery was operated at 52,000 rpm *in air* to attain the highest performance yet reported for *any known scanning system* (holographic or conventional): 20,000 elements/scan at 200 megapixels/sec. Augmenting the limited distribution government report (Bei 16) on this work which was concluded in 1972, results were presented at a conference in 1973 (Bei 4), coauthored by physicist Emile Darcy and engineer David Kleinschmitt. Physicist Andrew Dalisa provided valued early analysis and experiment. This milestone Holofacet scanner is now in the permanent collection of the Smithsonian Institution at the National Museum of American History in Washington, D.C.

During the May 1969 genesis of this system a method (Bei 18) utilizing solid transparent substrates also evolved. Following disclosure by this author on May 24, 1969, it was rendered by mechanical designer Al Bodner (AB) on May 29, 1969 as represented in Fig. A1.8. This implementation allows operation in

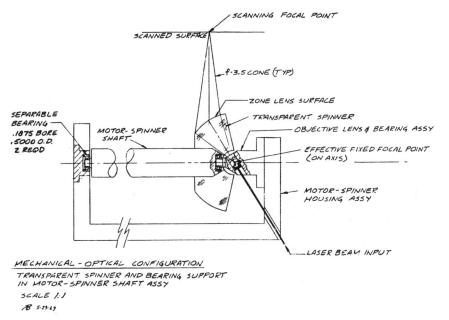

Figure A1.8. Holographic scanner disclosed May 44, 1969 utilizing solid transparent substrate, designed for high-speed stability with outboard bearing mounts (no cantilever) and extremely high resolution (scanning $f/3.5$ cone).

transmission with almost as high inertial stability as the above-described system, and substantially higher than those of the hollow shell techniques discussed subsequently. This method is described further in Section 4.2.1.3.1.

In March 1969 (Bei 15), predating the concave and convex spherical shapes noted above, consideration was given to the cylindrical shape for operation in transmission or reflection, to launch output beams normal to the axis. While operation in transmission poses no problem regarding reillumination with a valid conjugate (see Section 4.1.5), it was not pursued then because of excessive inertial stress upon the hollow cylinder in high-performance systems. It was not until October 1972 (Bei 17) that the illumination problem was resolved for operation in the inertially stable reflection mode, requiring axiconal optics to generate a line focus on axis for rigorous conjugate reference-reconstruction relationships in wide-aperture cylindrical and conic systems (see Fig. 4.18). An ultrahigh-performance design based on the solid cylindrical configuration is represented in Section 4.2.1.2.1. Subsequent systems developed for the much more relaxed business graphics applications are discussed here in Section A1.4 and in Section 4.2.1.3.2.

Returning to the disk configuration, an interesting U.S. patent issued to A.

Bramley on March 20, 1973 (Bra 2) which attempted to rectify the arcuate scan from such a diffractive scanner (illuminated normally) by adding a second one in tandem which operates to complement the bow generated by the first. While the effectiveness of detailed implementation is uncertain, the approach was novel in that the output beam traversing a pair of counterrotating (transmissive linear) gratings was intended to propagate in a plane (as a fan) about an undeflected axis which is parallel to the rotating axis, not normal to it. The approach is basically similar to that of a counterrotating pair of equal optical wedges which resolve into a resultant wedge which changes only in magnitude about a fixed axis.

Although this patent was filed in March 1970—long after the intense revitalization of holography—the author expressed no awareness of the option of forming his linear phase gratings interferometrically. Instead, the gratings were to be assembled (without adequate description) from parallel strips of miniature prism "blocks" (of rectangular or triangular cross section) deposited on an optical flat. The use of "conventional diffraction gratings" was noted, but the author deplored the extremely low first-order diffraction efficiency (of the presumably ruled types available then). Also disclosed (without sufficient support) was the idea of adjusting the phase delay of the periodic structure by compressing (or decompressing) the grating structure, which is sandwiched between two optical flats. This and other aspects of the patent augment its main coverage of scan straightening by means of counterrotation of tandem disk scanners, a concept subsequently analyzed and enhanced by Wyant (Wya) in 1975 and discussed here in Section 4.2.2.5.2.

Although not issued until March 5, 1974, a significant work of that period (filed December 15, 1971) is represented in a U.S. patent by J. W. Locke and D. Mills (Loc). This is an example of a reflective flat holographic scanner which, as expressed in the specification, was preferred over the transmissive disk described by Cindrich (Cin) because of its ease of stable support and control over the quality of only one surface rather than two surfaces and the intervening volume of a transmissive member. This work, assigned to the Communications Satellite Corp. (Comsat), was supported by the Canada Centre for Remote Sensing (CCRS) and was conducted in Canada by John Locke of the University of Toronto. It was a heroic effort, as perceived by this author during a visit to the CCRS in early 1975, intended for the rather challenging task of recording the NASA ERTS and Landsat images. Since an overilluminated radially symmetric disk-type scanner generates a circular locus (displaced from the disk and) concentric with the axis, a concentric cylindrical storage medium transported parallel to the axis will be recorded with a rectilinear raster. The experimental unit included *four* sets of 9-in. cupped film transports, each to be recorded with simultaneous full-color images. The disk was originally a glass (and later a metal) plate, metallized for reflectivity, coated by Kodak with silver halide emulsion and mounted to the spinner mandrel. The three hologram facets were to be exposed simultaneously per color, one for each of three sequential color exposures. (The discussion in Section 4.4.1 relates to some consequences of such exposure procedures.) During the development of this rather complex system, the holographic disk was replaced with a three-facet pyramidal polygon to align and set up instrumentation independently of the holographic process. Discussion and illustration of this system appear in Section 4.2.2.2.2.

Charles S. Ih was another member of the CBS Laboratories' staff who was active during the early stages of holographic scanner development. He and one of our research team members, David Kleinschmitt, contributed significantly to the wideband Holofacet evaluations by conducting the follow-up effort for the Air Force, which resulted in additional film recording tests concluded in December 1973 (Ih 3). Stabilizing the exposure and test facilities, and incorporating a broadband electrooptic modular and driver subsystem (Fig. 4.13) they formed high-quality holograms on the spherical substrate. Dynamic tests at 52,000 rpm *in air* produced 100 lp/mm recordings at 33% MTF on type E.K. 3414 photographic film. The corresponding video bandwidth was 100 MHz.

The work for which Charlie Ih is most noted is the introduction of the auxiliary reflector upon a reflective disk-like spinner to transform the scanned beam to a locus normal to the rotating axis. This was first presented almost simultaneously by Ih at the OSA Annual meeting (paper ThC October 21, 1974) and by L. Beiser describing the work of Ih at the 1974 E-O Systems Design Conference and published in the proceedings (Bei 7). Ih's patent (Ih 1) was filed on October 16, 1974 and on issue April 27, 1976, was assigned to Epsco, Inc., the organization that continued with some of the laser scanning work of CBS Laboratories as Epsco Labs. While many variations of the auxilliary reflector have appeared since, the most descriptive first major publication (Ih 2) was somewhat delayed until submitted in March 1977 under the auspices of the University of Delaware, where Ih relocated and continued his work. The auxiliary reflector is discussed in Section 4.2.2.3.

Some of the most innovative variations to reflective holographic scanning were created by David Kleinschmitt (Kle 1, Kle 2), described in part in Section 4.1.8.2 for his ellipsoidal and parabolic substrates and illustrated in Fig. 4.9 for his disk substrate. The disk configuration also appeared as Fig. 2 in our 1973 joint publication on the Holofacet scanner (Bei 4). It was disclosed on December 3, 1973 (Bei 5) and after typical effort in filing for patent, was rejected on May 1975 because of prior art of Locke (Loc) and Beiser (Bei 1), both described earlier. Another technique by Kleinschmitt reduces the facet-accuracy requirements of a conventional-looking pyramidal polygon (Kle 1). Starting with one having pyramidal angles of approx. 60° (not critical, or required to be consistent) and applying a controlled set of holograms thereto, angularly precise outputs will be reconstructed normal to the axis when illuminated by a collimated beam parallel to the axis. The key to this achievement is that the reference beam is trimmed in angle prior to each holographic exposure (with the object beam fixed and directed normal to the axis) such that its reflection propagates parallel to the axis. This nulls each facet angular error such that the reconstruction beam (its conjugate) will form the diffracted output beam at precisely the original (correct) angle.

As delayed for Ih (discussed earlier), these innovations by Kleinschmitt did not surface beyond the formative levels because of organizational transitions from CBS Laboratories. Other major contributors to holographic scanning from that pioneering team were physicists Emile Darcy (Bei 16, Bei 4) and Andrew Dalisa (Bei 15, Dal), for whom we respectfully reflect our recognition and sincere appreciation for their dynamic and creative support.

A1.4. MOTIVATION OF BUSINESS GRAPHICS

Most of the earlier work was directed toward very high performance imaging, motivated by government agencies rather than by private industry. While the potential need for business graphics (scanning/digitizing and printing) was anticipated by earlier investigators (e.g., Fig. 4.24 and associated discussion), it was not until 1975 that the leaders in the business machine field expressed pursuit of holographic scanning to match their more moderate resolutions and speeds. The first announced activity appeared almost simultaneously from IBM and Xerox. IBM's work by Pole and Wallenmann (Pol 1) is reviewed in Section 4.2.1.3.2. The corresponding patent by Pole and Werlich (Pol 2) was filed May 1977 following co-pending application filing of December 1975. It was issued September 1978 and includes coverage for certain cylindrical and disk-type transmissive scanners. It is limited to those that utilize retrocollection: backscatter from radiation from an opaque document which is rediffracted by the hologram for signal detection (see Fig. 4.23 and its discussion). This may reflect the same frustration highlighted earlier of being anticipated by works related to Baird for his diffractive hollow cylinder of 1934 and by Jenkins for his lens-disk of 1928, although the patent does not refer to this earlier work. Another IBM patent (Whi) issued a month later, incorporates the transmissive cylinder disclosed by Pole on August 1974 in IBM Research Paper RJ1423, and does not require signal detection. However, the reillumination is reoriented to be radially asymmetric to tend to linearize the scan, departing thereby from the prior art, which was generally radially symmetric.

The first Xerox effort announced during this period involved investigating and testing the use of computer-generated diffractive elements on translating or rotating substrates. In 1975, Olof Bryngdahl (Bry 1) published his first work (received by *Optics Communications* June 1975). It was followed (in January 1976) by a more comprehensive presentation coauthored by Wai-Hon Lee (Bry 2) (received by *Applied Optics* August 1975). As with IBM's proprietary coverage, the patent for this process (Bry 3) was issued substantially later (August 1978), for it was filed February 1977 as a continuation of their original filing of June 1975. The purpose of computer generating the holograms was to impart spatial control over the scanning process in one, two, or three dimensions which would otherwise be extremely difficult (if not impossible) to accomplish with interferometrically generated gratings. One potentially useful application would be the dynamic correction of focal length and beam displacement as a function of scan angle, to form a more linear and more well-focused flat-field scan function without the use of flat-field lenses. A discussion of computer generation of holograms and their rotational scanning appears in Section 4.5. Their translational scanning discussion appears in Section 4.3.

Shortly following this work by Xerox in its Palo Alto Research Center, several contributions were expressed and expanded by Charles Kramer (at the Xerox Wilson Center near Rochester, New York) which matured into a prominent direction in holographic scanning. Long seeking a solution to the disk wobble problems (which cause the scanned beam to be misplaced), in May 1976, Kramer filed for, and in January 1978 obtained, patent coverage (Kra 1) for a mechanical means for stabilizing a rotating holographic disk. By mounting the disk flexibly to the shaft through an elastomer, the radial inertial forces due to disk rotation tend to align it in a plane normal to the axis. This general approach was the subject of significant attention by Kramer and in a 1979 presentation (Kra 2), three additional techniques were described which utilize aligning methods which are free of the residual counterforce of elastomers that limit perfect orientation.

Concurrent with this work in mechanical self-alignment was a search for a more elegant "automatic" solution. Two months following the filing of the elastomeric approach, Kramer filed for coverage of a fundamental departure from prior holographic disk utilization which nulls the effect of disk wobble by setting the illumination and diffraction angles to approximate equality about the disk. This patent (Kra 3), issued in December 1980, retained the concept of radial symmetry and the implicit need for additional correction to straighten the bow due to nonnormal propagation of the scanned beam. Thus in a patent filed later but issued almost concurrently (Kra 5) appears the incorporation of an auxiliary concave reflector similar to that described by Ih and discussed here and in Section 4.2.2.3. This unbows the scan line and at the same time becomes part of the wobble-correction system such that both wobble and bow are nulled. Residual difficulties remained, however: The concave reflector is extremely sensitive to centration errors and complicates system alignment for diffraction-limited performance. Further, because the holographic facets provide optical power, there is no simple provision for stigmatic imaging following wavelength shift between exposure and reconstruction (see Section 5.4).

With continued attention to these problems, on May 31, 1979 the filing for the composite solution, which is now identified with Kramer's work, emerged. This patent (Kra 7), issued September 15, 1981 [shortly after the first publication of its characteristics (Kra 6)], allowed for elimination of the spherical reflector and formation of linear plane gratings (no optical power). Under the conditions of setting the input and diffraction angles at approximately 45°, the scanned output beam traverses a path that remains essentially in a plane over a sufficiently wide range to be practical. Wobble nulling is retained and correction for wavelength shift is allowed. Although the technique is not without critical control requirements (covered in Sections 5.2 and 5.3), these searching transformations yielded one of the most significant contributions to holographic scanning, and an impetus for continued advancement.

Quite remarkably, a parallel development which complements the work by Kramer and draws similar, although more restricted conclusions, emerged from the Soviet Union (Ant). This elegant work by Antipin and Kiselev was published in June 1979, during the interval between filing and release of Kramer's publications on reduction of the effects of disk wobble. The analytic approaches are sufficiently different to support the position that they were created independently.

Recalling the two phases of development by Kramer, where the first yielded the equal-angle criterion and the second concluded the use of linear gratings and approximated straight scan on a plane surface, his relatively early filing (July 1976) for the first criterion predates publication by Antipin and Kiselev by three years and indicates a likely priority by Kramer for this important factor. Although the second conclusion exhibits more parallel timing, without further basis for interactions between these two remote sources, we conclude that they were indeed independent, with at least some temporal priority by Kramer.

The Soviet publication is available in very few libraries. Because of its importance in analytic orientation and the need for cohesive interpretation, in Appendix 2 we provide an annotated translation of this work. To clarify another perception of the problem and appreciate its solution, in Appendix 3 an annotated abstract of Kramer's patent concentrating on scan bow minimization, is provided. In Section 4.2.2.5 further representation of both works is presented.

A1.5. LINEAR TRANSLATION AND STEPPING

The activity discussed earlier regarding scanning computer-generated holograms (Bry 1, Bry 2, Bry 3) was applied mainly to rotating systems, seeking to generate linearized scan functions which emulate those of more familiar scanners, as well as forming general scan functions having no conventional counterpart. In 1977, Gerbig (Ger 1) divided the computer-generated holograms into subsets or arrays for discrete positioning serially, parallel-serially, and in tandem to achieve two- and three-dimensional scan. This work was continued jointly with S. Case, utilizing interferometrically generated holographic elements, reported starting in 1980 (Cas 2). The potential advantages of such manipulation relate to the formation of general and arbitrary scan functions in x, y, and z. The limitations, as appreciated by the authors, relate to the seep-wise nature of the image points or to its consequential aberration if continuity is attempted. To achieve high diffracted output purity with continuous scan, computer-generated holograms were operated in a one-dimensional form (Ger 2) and coupled through anamorphic optics to achieve two- and three-dimensional image point placement. Translational positioning holographic scanning is discussed in Section 4.3 and as with the related work of Bryngdahl and Lee discussed earlier, computer-generated holographic scan is covered in Section 4.5.

A1.6. CONTINUATION OF DEVELOPMENT

Continuing technological development may be placed in to the category of recent history, represented in the body of this volume, where many of the contributions appearing in this appendix are discussed more completely. Also, the next two appendixes detail some principal novelties developed along this innovative route. Although not motivated toward completeness, it is believed that this combination of sources adds a unique temporal perspective to the principal mission of documenting the technology of holographic scanning.

Appendix 2

Annotated Translation of Laser Beam Deflector Utilizing Transmission Holograms*

INTRODUCTION

This annotated translation of an almost inaccessible publication provides a unique resource for analysis and reference. The original work was published in Russian in 1979. It remained obscure until 1982, when a first translation was conducted by Sigfried H. Mohr and his associates Alex Miroshnichenko and Michail Veprinsky. Although its contents appeared cohesive upon casual reading, critical analysis and validation were severely frustrated by the many and sometimes crucial typographical errors in the original, by excessively compact analytic expression by the authors, and by the translational rigidity of technical and professional jargon. We hope to have removed these limitations in this annotated translation.

Liberty was taken here in transforming literally translated words and phrases into expressions more familiar to the technical community. My annotations are introduced (in italic type) to clarify statements or to explain editorial action. Figure 2 is redrawn as Fig. 2.2' to render its role more effective and to include some notations. Finally, gross voids in analytic continuity are here filled to provide more confidence in its content and access to its derivation.

To minimize disruption of logical flow in the original work, major annotations appear at the end of the Appendix. Superscript numerals added to the main section key the reader to consult these explanations and derivations, given

*From Technika Kino I Televideniya (Motion Picture and Television Engineering), Monthly Scientific Engineering Journal of the USSR Government Committee for Cinematography, No. 6 (pp. 43–45), Leningrad Institute of Motion Picture Engineers. (June 1979).

191

in the "Keyed Annotations" at the end. Also, the original nomenclature and notation are retained. A table at the end of this appendix provides some of the notation equivalents to those in the body of this volume.

LASER BEAM DEFLECTOR UTILIZING TRANSMISSION HOLOGRAMS
by M. V. Antipin and N. G. Kiselev

One of the best ways of transferring an image from magnetic tape onto motion picture film is the method based on using the laser as a light source for opto-mechanical line formation [1]. The most difficult element for realization of the above method is the subsystem for line scan which uses a high-speed mirror drum (polygon). The generation of such a drum requires achievement of accuracies to seconds of arc of the angular position of the various facets and is quite a difficult engineering task. This motivates us to look for a simpler realization of this subsystem.

A possible approach for realizing the line scan function is the so-called "holographic deflector" (HD), which consists of an array of identical holograms which are disposed on the surface of the rotating body.

The principle of operation of the various HD configurations is described in a series of papers [2–7]. From an analysis of these papers it follows that one of the first tasks is the selection of an HD arrangement that ensures linearity of the movement of the scanning light spot along with the absence of line curvature. In particular, the HD suggested in [2, 3, 7] forms a curved line, and because of this, the useful angle of deflection in light of the above-mentioned requirements is not large. Some suggested devices [4, 6] do not have this problem, but from the construction point of view they are relatively complex because nonflat surfaces are used as the hologram substrate.

One notes that the basically advantageous characteristic of the HD is the potentially high accuracy of the spatial deflection of a laser beam which is achieved during its formation by simpler techniques than in the case of the traditional mirrored polygon. The holograms are formed by noncontact methods which do not deform the substrate, and high spatial accuracy of deflection is assured by precise rotation relative to the axis, established in the exposure stage, and the following quality chemical processing of the photo-sensitive emulsion.

In contrast to the prior art noted above, a simpler construction method for the HD is provided in this article, realized on a flat substrate and utilizing transmission holograms. Despite its simplicity, the suggested configuration yields deflection nonlinearity of less than 1% within a beam deflection angle of 32°, along with practical absence of line curvature.

THEORETICAL ANALYSIS

The arrangement of the line scan subsystem of the suggested configuration for an HD is shown in Fig. 1 [Fig. A2.1]. A flat transparent substrate (2) with holograms (8) is fixed to the shaft of the electric motor (1). A collimated beam (3) is incident on the substrate and goes through one hologram (8) and forms a collimated diffracted beam (4), focused by a lens (5) in the plane of the recording photo-material (e.g., motion picture film). While the hologram rotates, a line of a frame (7) is formed in the film plane. In order not to have a blanking interval, the illuminating beam covers two adjacent holograms on the surface of the substrate. A smooth film transport provides formation of a two-dimensional raster.

Let us look at the principle of operation of the HD with the help of Fig. 2 [Fig. A2.2 and Fig. A2.2']. Here the directions of the incident waves are indicated by their position vectors. Let two plane waves: the object wave $\overline{O}$ and the reference wave $\overline{P}$, form the hologram in the form of a sinusoidal grating in the xoy plane where the recording material is located. After chemical processing the hologram is returned to its original position and it is illuminated by wave $\overline{C}$.

Note: Reconstruction wave $\overline{C}$ usually complements the reference wave $\overline{P}$ rather than the object wave $\overline{O}$. This reversed procedure has no effect on the results. (Russian letter "P," pronounced "R," denotes "reference".) Vector bar notations in this paragraph rectify apparent typos in the original. Also, coordinate notations are lowercase per Fig. 2 [Fig. A2.2].

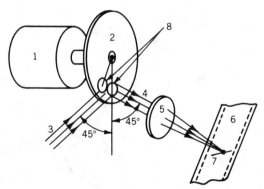

Figure A2.1. Arrangement for line recording: 1, electric motor; 2, support plate with holograms; 3 and 4, recreated incident and diffracted beams; 5, focusing lens; 6, photographic recording material; 7, line formed by the holographic deflector.

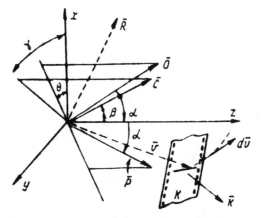

Figure A2.2. (Fig. 2 of original work.) Defining coordinates of directional position vectors of the diffracted beam: O, P, C are directional position vectors of the object, reference, and reconstruction waves; R, V are directional position vectors of the real and virtual diffracted waves; K is directional position vector of a normal to the plane K of the photo-material; dv is a vector tangential to the trajectory of the displacement of the end of the position vector. γ is the angle of plane of $\overline{C}$ with respect to the x axis.

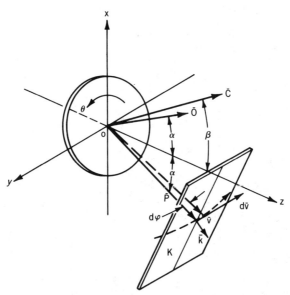

Figure A2.2′. Fig. A2.2 clarified and adjusted for perspective. $\overline{C}, \overline{O}$ and $\overline{P}$ in xoz plane ($\gamma = 0$). R deleted and $o\&d\varphi$ added.

194

It is known [8] that the complex amplitude of the recreated wave is made up of a real part H_R and of a virtual part H_V which can be defined as follows:

$$H_R = \overline{C}\,\overline{O}\,\overline{P}* \tag{1}$$

$$H_V = \overline{C}\,\overline{O}*\overline{P}, \tag{2}$$

where $\overline{C}$, $\overline{O}$, $\overline{P}$ are complex amplitudes of the illuminating, object, and reference waves, and * indicates the complex conjugate value. Taking into consideration that the amplitudes of all three waves are the same in time in the plane *xoy*, one can consider only the phase contributions of expression (1) and (2) and then (2) takes the form

$$H_V = \exp\left\{ j_o \frac{2\pi}{\lambda} \left[\Delta_c(x, y) - \Delta_o(x, y) + \Delta_p(x, y) \right] \right\}. \tag{3}$$

where $\Delta_o(x, y)$, $\Delta_p(x, y)$, $\Delta_c(x, y)$ are the differences of the displacement of the different rays of parallel beams to the plane *xoy* of the object, reference, and illuminating wave, and therein: λ is the wavelength of the incident wave and j_o is the imaginary unit.

Let us assume that for each of the three waves the following is known, namely the directional cosines of the normals to the plane wave:

$$\overline{O} = O_x \overline{i} + O_y \overline{j} + O_z \overline{k};$$

$$\overline{P} = P_x \overline{i} + P_y \overline{j} + P_z \overline{k};$$

$$\overline{C} = C_x \overline{i} + C_y \overline{j} + C_z \overline{k}.$$

By using the nomenclature from Fig. 2, let us write the following:

$$O_x = \sin \alpha \cos \Theta; \quad P_x = -\sin \alpha \cos \Theta; \quad C_x = \sin \beta \cos \gamma;$$

$$O_y = \sin \alpha \sin \Theta; \quad P_y = -\sin \alpha \sin \Theta; \quad C_y = \sin \beta \sin \gamma;$$

$$O_z = \cos \alpha; \quad P_z = \cos \alpha; \quad C_z = \cos \beta.$$

Angle Θ, which is taken in the positive direction from the axis *ox*, is the angle of rotation of the hologram around the axis *oz*. For $\gamma = 0$ *(reconstruction vector $\overline{C}$ in xoz plane)*, $C_x = \sin \beta$ and $C_y = 0$. The difference in movement of any ray of a parallel beam (e.g., object wave) incident to a point with coordinates x, y in the plane xoy compared with a ray that was incident to the origin of the coordinates is defined as follows:

$$\Delta_o(x, y) = xO_x + yO_y. \tag{4}$$

Expressing similarly the difference in movement for the reference and the reconstruction beams and using Eq. (3), we obtain the direction cosines v_x, v_y, v_z of the virtual wave recreated from the hologram, as follows:

$$v_x = \sin \beta \cos \gamma - 2 \sin \alpha \cos \Theta;$$

$$v_y = \sin \beta \sin \gamma - 2 \sin \alpha \sin \Theta; \tag{5}^{1*}$$

$$v_z = \left(1 - v_x^2 - v_y^2\right)^{1/2}$$

$$= \left[\cos^2\beta + 4 \sin \beta \sin \alpha \cos (\Theta - \gamma) - 4 \sin^2\alpha\right]^{1/2}.$$

From the manner in which v_z was determined, it follows that the variables α, β, γ, and Θ are to satisfy the following condition:

$$v_x^2 + v_y^2 = \sin^2\beta - 4 \sin \alpha \sin \beta \cos (\Theta - \gamma) + 4 \sin^2\alpha \leqslant 1. \tag{6}^2$$

Equation (5), along with condition (6), yield the coordinates of the diffracting vector on the hologram wave.

Defining the Angular Deflection of the Diffracted Wave with the Rotation of the Holographic Deflector, HD

Let us form the differential $d\bar{v}$ of the radius vector $\bar{v}$ of the diffracted wave which is directed tangentially to the trajectory of the moving end of the position vector $\bar{v}$, and its absolute value is the elementary increase of this position vector:

$$d\bar{v} = d\Theta(v_x'\bar{i} + v_y'\bar{j} + v_z'\bar{k}), \tag{7}^3$$

where v_x', v_y', v_z' are the derivatives of the coordinates of the position vector $\bar{v}$. The angular deflection $d\varphi$ of the diffracted beam [see Fig. 2' (Fig. 4.34)] when the hologram is rotated through angle $d\Theta$ is equal to the ratio of the absolute values of vectors $d\bar{v}$ and $\bar{v}$, i.e.,

$$d\varphi = \frac{d\bar{v}}{\bar{v}} = d\Theta\left[(v_x')^2 + (v_y')^2 + (v_z')^2\right]^{1/2}. \tag{8}$$

*Refer to "Keyed Annotations."

Furthermore, to simplify the analysis, let us assume that the illuminating beam is in the same plane with the object and reference wave at $\Theta = 0$. That is, $\gamma = 0$ (per Fig. 2'). Upon substitution, Eq. (8) becomes

$$t = \frac{d\varphi}{d\Theta} = \sin\alpha\left[4 + \frac{\sin^2\beta\,\sin^2\Theta}{\cos^2\beta + 4\sin\alpha\,\sin\beta\,\cos\Theta - 4\sin^2\alpha}\right]^{1/2}. \quad (9)$$

From here it follows that the linearity of the angular deflection of the diffracted beam is valid only for the condition that $\beta = 0$, i.e., when the illuminating beam is incident normal to the hologram plane. In other cases nonlinearities occur, and when tight tolerances are expected from this parameter, one has to minimize the deflection angle.

If the illuminating beam is coincident with the object beam at $\Theta = 0$, then $\alpha = \beta$ and Eq. (9) reduces to

$$t = \frac{d\varphi}{d\Theta} = \sin\alpha[4 + \tan^2\alpha\,\sin^2\Theta]^{1/2} \quad (9a)^4$$

With $\alpha = 45°$ the nonlinearity σ of the angle of deflection of the diffracted wave is calculated from the following formula:

$$\sigma = \frac{t - t_{\Theta=0}}{t_{\Theta=0}} \times 100\% \quad (9b)^5$$

and does not exceed 1% with the angle of rotation of the hologram $|\Theta| \leqslant 11.5°$ and with the full angle of the deflection of the beam $|\varphi| \cong |2\Theta\,\sin\alpha| \leqslant 16°$, i.e., for an angular width of the line of 32°.

Determination of the Conditions under Which the Curvature of the Line is a Minimum

To solve this problem let us define the equation of the line that is formed by intersecting the diffracted beam with some plane K that is at a distance 1 from the origin of the coordinates and is parallel to the plane of the photo-material.

The coordinates of the point of intersection (x, y, z) of the diffracted beam with the plane K appears as follows:

$$x = \frac{v_x}{\cos\tau}; \quad y = \frac{v_y}{\cos\tau}; \quad z = \frac{v_z}{\cos\tau}. \quad (10)^6$$

where $\cos \tau = k_x v_x + k_y v_y + k_z v_z$ and k_x, k_y, k_z are the direction cosines of the normal to the plane of the recording material.

Let plane K be parallel to axis oy (see Figs. A2.2 and A2.2'). Then $k_x = \sin \chi$, $k_y = 0$, $k_z = \cos \chi$, and $\chi = \Theta$.[7] The curvature of the line will be a minimum if the change in coordinate z with a change of angle Θ is equal to zero, i.e.,

$$\frac{dz}{d\Theta} = \frac{k_x}{(k_x v_x + k_z v_z)^2} (v_x v_z' - v_z v_x') = 0. \qquad (11)[7]$$

Condition (11) is met when

$$v_x v_z' - v_z v_x' = 0. \qquad (12)$$

After substitution of the actual values for v_x and v_z from (5) into (12), we obtain

$$-\frac{v_y}{v_z} (1 - 4 \sin^2\alpha + 2 \sin \alpha \sin \beta \cos \Theta) = 0. \qquad (13)[8]$$

This establishes that the curvature of the line is a minimum when the relationship between α, β, Θ is as follows:

$$\sin \beta = \frac{1}{\cos \Theta} \left(2 \sin \alpha - \frac{1}{2 \sin \alpha} \right). \qquad (14)[9]$$

Taking into consideration that $2 \sin \alpha = \lambda/d$, where d is the diffraction grating constant, (14) takes the following form:

$$\sin \beta = \frac{1}{\cos \Theta} \left(\frac{\lambda^2 - d^2}{\lambda d} \right). \qquad (15)[10]$$

From (15) can be seen that there are several variations that satisfy condition (14)–(15). For a number of reasons the most attractive is the symmetrical variation when $\alpha = \beta$ at $\Theta = 0$. The expression (14)–(15) in this case determines that

$$\alpha = \beta = 45°; \quad d = \frac{\lambda}{\sqrt{2}}. \qquad (16)$$

The curvature of the line is determined as the ratio of the deviation of the trajectory movement from a straight line $\Lambda = z_o - z$ (where $z = z_o$ when $\Theta = 0$)

to the angular width of the line $2y$, with $y = v_y/\cos \tau$. In this case it does not exceed 0.0006 in the region of the movement of the HD of $\pm 8°$ or a deflection angle of the diffracted beam of $\pm 12.7°$.

Experimental Study of the HD Model

It is the goal of the experimental test to show that an HD can be realized on the basis of the theoretical statements above. The optical arrangement for producing the HD is shown in Fig. 3 [Fig. A2.3]. The beam from the He-Ne laser LG-38 is enlarged by the telescope system 2 and was directed onto the surface of the supporting plate 3 with the photosensitive emulsion, forming the reference wave. Mirror 4, also located in the collimated beam, formed the object plane wave. The flat support plate was held firmly on the shaft of the electric motor (5) with a titanium adapter. The motor shaft was connected to the indexing mechanism by means of a coupling (7), ensuring the accuracy of shaft rotation to ± 1 minute of arc.

The holograms are recorded on a circular path with an average radius of 30 mm. The number of holograms was equal to 18, such that the dimensions of a separate hologram were equal to 10×10 mm. The parameters of the recording arrangement were as follows: $\alpha = \beta = 45°$ (from here the line frequency $\nu = \sqrt{2}/\lambda = 2240 \ 1/\text{mm}$); the angular increase in the velocity of deflection, t, calculated from equation (9) then yields $t = d\varphi/d\Theta = 2 \sin \alpha \simeq 1.41$. Thus the diffracted beam moves 1.41 times faster than the hologram itself. The theoretical number of elements in the line (*resolution*) corresponds to the ratio of the full angular deflection $2\Theta \sin \alpha$ to the (*diffractive*) angular spread of the beam λ/l_g (where l_g is the active hologram size) equals 5550.

For the photosensitive emulsion, high-resolution silver halides (product emulsion LO1-2) were used (developed at LIKI by N. N. Jaroslawski). Pro-

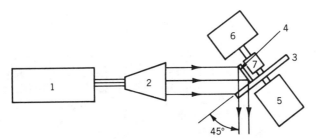

Figure A2.3. Arrangement for exposing the HD: (1) He-Ne laser LG-38; (2) telescope system; (3) substrate with photosensitive emulsion; (4) mirror; (5) Electric motor; (6) indexing arrangement to obtain precise angle; (7) shaft coupling.

cessing of these photoemulsions was done in the developer GP-2. The diffraction efficiency of emulsion LOI-2 was equal to 25%; it was a little lower for emulsion ÁVR-LIKI because the developer GP-2 is not optimal for it.

The quality of the reconstructed wavefront was evaluated by visual inspection of the form and the size of the resulting spot formed in the focal plane of a lens of a high-quality telescope. Typically, the central spot and the rings surrounding it were slightly deformed. This is apparently related to the small-scale deformations on the surface of the gelatin. One can determine the theoretical diffracted spot size from the formula

$$\delta = \frac{2\lambda f}{l_g},$$

where f is the focal length of the lens used, and $l_g = 7$ mm. For $f = 400$ mm and $\lambda = 0.63$ μm we obtain $\delta = 0.072$ mm. While the actual size of the spot averaged twice as large, one can agree that this is acceptable because special requirements in the quality of chemical processing were not demanded.

The error of the location of the line in a direction perpendicular to the trajectory of the scanning spot was measured by goniometer GS-5. Noncoincidence of lines from the formed holograms was 2 minutes of arc. It was observed that a characteristic displacement of the lines took place from hologram to hologram, caused by the segment configuration of the hologram substrate and by some variation in the axis of rotation.

Visual observations of the trajectory of the movement of the scanning spot during the rotation of the hologram showed that when the reconstruction beam was incident at an angle of 45° to the substrate and for small variations from this value, curvature of the line was not observable despite the quite large angular deflection. However, when this (incident) angle deviates by 5 to 10° in one of the directions, curvature of the line can be observed, but depending on the direction of the deviation, it has a different sign of curvature.

Conclusion

The feasibility of a flat-field laser beam deflector was shown theoretically and experimentally to exhibit practically no line curvature. For practical utility, it is necessary for the diffraction efficiency of the holographic deflector (HD) to be at least 50% and to eliminate the wavefront distortion of the diffracted beam.

REFERENCES

[1] Antipin, M. V., and Kiselev, N. G. Technology on development of laser recording systems, *Tech. Kino Telev.*, 1977, No. 1, p. 71–79.

[2] Cindrich, I. (*Cin*).* Image scanning by rotation of a hologram. *Appl. Opt.*, 1967, No. 9, p. 1531–1534.

[3] McMahon, D. H., Franklin, A. R., and Thaxter, I. B. (*McM 1*). Light beam deflection using holographic scanning techniques. *Appl. Opt.*, 1969, No. 2, p. 399–402.

[4] Pole, R. V., and Wollenmann, H. P. (*Pol 1*). Holographic laser beam deflector. *Appl. Opt.*, 1975, No. 4, p. 976–980.

[5] Bryndahl, O., and Lee, W. H. (*Bry 2*). Laser beam scanning using computer-generated hologram. *Appl. Opt.*, 1976, No. 1, p. 183–194.

[6] Beiser, L. (*Bei 6*). Laser scanning systems. *Laser Applications*, Vol. 2. New York: Academic Press, 1974, p. 152–153.

[7] Ih, C. S. (*Ih 2*). Holographic laser beam scanners utilizing an auxiliary reflector. *Appl. Opt.*, 1977, No. 8, p. 2137–2146.

[8] Collier, R. (*Col*). Burckhardt C., and Lin L. *Optical Holography*, 2nd ed. New York: Academic Press, 1971; Moscow: Peace, 1973, p. 86, 69.

KEYED ANNOTATIONS

1. (*Eq. 5*): $\left.\begin{array}{l} v_x = C_x - O_x + P_x = \sin \beta - 2 \sin \alpha \cos \Theta \\ v_y = C_y - O_y + P_y = -2 \sin \alpha \sin \Theta \end{array}\right\}$ for $\gamma = 0$.

2. (*Eq. 6*): $v_x^2 + v_y^2 = 1 - v_z^2$, and expand v_z^2.

3. (*Eq. 7*): $v_x' \equiv dv_x/d\Theta$, etc.

4. (*Eq. 9a*): added "(9a)." When $\alpha = \beta$, fraction in (9) reduces to

$$\frac{\tan^2\alpha \, \sin^2\Theta}{1 + 4 \tan^2\alpha(\cos \Theta - 1)}$$

whence $t(\Theta \approx 0) = 2 \sin \alpha$.

5. (*Eq. 9b*): added "(9b)."

6. (*Eq. 10*): typographical errors in original y term.

7. (*Eq. 11*): Introductory statement exhibits disruptive typographical error in original, showing $\chi = 0$ rather than the correct $\chi = \Theta$. Derivation procedure as follows:

Per Eq. (10) and cos τ conditions,

$$z = \frac{v_z}{(\sin \Theta)v_x + (\cos \Theta)v_z} \tag{a}$$

*Note: Refs. in ital. in parens give location in this volume's reference list.

For $v \neq f(\Theta)$, solving for $dz/d\Theta$ yields, for $v_z \neq 0$,

$$(\sin \Theta)v_z - (\cos \Theta)v_x = 0 \qquad \text{(b)}$$

For $v = f(\Theta)$, then

$$z = \frac{v_z(\Theta)}{(\sin \Theta)\, v_x(\Theta) + (\cos \Theta)\, v_z(\Theta)} \qquad \text{(c)}$$

$$\frac{dz}{d\Theta} = \frac{(\cos \tau)v_z' - v_z\{[(\sin \Theta)v_x' + v_x(\cos \Theta)] + [(\cos \Theta)v_z' - v_z(\sin \Theta)]\}}{\cos^2 \tau} \qquad \text{(d)}$$

with $\cos \tau = (\sin \Theta)v_x + (\cos \Theta)v_z$, this reduces to

$$\frac{dz}{d\Theta} = \frac{\sin \Theta (v_x v_z' - v_z v_x') + v_z[(\sin \Theta)v_z - (\cos \Theta)v_x]}{\cos^2 \tau}$$
$$= 0 \qquad \text{(e)}$$

Letting $\sin \Theta = k_x$, $\cos \Theta = k_z$, $\cos \tau = k_x v_x + k_z v_z$ and the right term in the numerator $= 0$ for $v_x \neq f(\Theta)$ per Eq. (b) above, Eq. (e) reduces to Eq. (11).

8. (*Eq. 13*):   Per Eq. (5) at $\gamma = 0$,

$$v_x = \sin \beta - 2 \sin \alpha \cos \Theta$$
$$v_y = -2 \sin \alpha \sin \Theta \qquad \text{(a)}$$
$$v_z = (1 - v_x^2 - v_y^2)^{1/2}$$

Then

$$v_x' = 2 \sin \alpha \sin \Theta = -v_y$$
$$v_y' = 2 \sin \alpha \cos \Theta = \sin \beta - v_x \qquad \text{(b)}$$
$$v_z' = (-v_x v_x' - v_y v_y')/v_z$$

Substituting into Eq. (12) yields

$$v_x v_z' - v_z v_x' = (\sin \beta - 2 \sin \alpha \cos \Theta)(-v_x v_x' - v_y v_y')/v_z$$
$$- (1 - v_x^2 - v_y^2)^{1/2}(2 \sin \alpha \sin \Theta) \qquad (c)$$

Expanding terms gives

$$v_x v_x' = 2 \sin \alpha \sin \beta \sin \Theta - 4 \sin^2\alpha \sin \Theta \cos \Theta$$
$$v_y v_y' = (-2 \sin \alpha \sin \Theta)(2 \sin \alpha \cos \Theta)$$
$$= -4 \sin^2\alpha \sin \theta \cos \Theta$$

Therefore,

$$v_x v_x' - v_y v_y' = 2 \sin \alpha \sin \beta \sin \Theta \qquad (d)$$

Substituting Eq. (d) into Eq. (c) yields

$$v_x v_z' - v_z v_x' = -\frac{v_y}{v_z}(v_z^2 + \sin^2 \beta - 2 \sin \alpha \sin \beta \cos \Theta)$$
$$(e)$$

Setting

$$v_z^2 + \sin^2\beta = 1 - v_x^2 - v_y^2 + \sin^2\beta \qquad (f)$$

when expanded and simplified reduces to

$$v_z^2 + \sin^2\beta = 1 - 4 \sin^2\alpha + 4 \sin \alpha \sin \beta \cos \Theta \qquad (g)$$

which when substituted into Eq. (e) yields Eq. (13).

9. (*Eq. 14*): Since $v_y/v_z \neq 0$ in Eq. (13), then

$$2 \sin \alpha \sin \beta \cos \Theta = 4 \sin^2\alpha - 1, \quad \text{etc.}$$

10. (*Eq. 15*): $\text{Sin } \beta = (1/\cos \Theta)(\lambda/d - d/\lambda)$ corresponds to Eq. (18*a*) in Appendix 3.

Some Nomenclature Equivalents

Appendix 2	Body of This Volume
$-y$	x
x	y
$|\bar{P}|$	R_s
Θ	Φ
φ	Θ
l_g	D
t	m

Appendix 3

Scan-Line Bow Minimization*

INTRODUCTION

This annotated rendering of an innovation in the field of holographic scanning offers enhanced understanding of the problem and the analytic perception directed toward its solution. The problem is the quest for a method of unbowing the scan line (upon a flat surface) projected by a disk-like scanner. The same problem is addressed in Appendix 2. The different approaches taken (almost simultaneously) are noteworthy in their contrasting analytic techniques.

Although this document retains integrity with almost verbatim abstract of the original, a significant amount has been augmented. Most notably, Fig. A3.5 here provides many clarifying additions, and key derivations help validate the work through invocation of some subtle relationships. Some of the explanatory material, notably that which yielded Eq. (11), is derived from private communication with Charles Kramer.

The original published work (U.S. Patent 4,289,371) provides additional considerations which are covered elsewhere in this volume. This abstract concentrates on, and where warranted, expands on the fundamental transformation of an arcuate scan to a straightened line. To minimize disruption of logical flow appearing in the original work, departures in the text are noted in italics, and superscript numerals key the explanations and descriptions which are added in the "Keyed Annotations" section at the end of this appendix.

For reference with the patent, the section entitled "Scan Line Bow Minimization" starts in the vicinity of col. 4, line 46 of the patent, paraphrasing and augmenting the descriptions to introduce Eqs. (3). Associated Figs. A3.3, A3.5, A3.6, and A3.7 have in some instances been annotated and clarified.

*Annotated Abstract from U.S. Patent 4,289,371 by Charles J. Kramer, "Optical Scanner Using Plane Linear Diffraction Gratings on a Rotating Spinner," filed May 31, 1979; issued September 15, 1981.

SCAN-LINE BOW MINIMIZATION

Figure 3 *[Fig. A3.3]* shows a tranmissive disk-like spinner located in an *X-Y* plane and rotatable about a *Z*-axis. A reconstruction wavefront (Input Beam) illuminates a plane linear diffraction grating (PLDG) facet (one of several disposed nonoverlapping about the periphery) at an angle of incidence Θ_i, and is diffracted at angle of diffraction Θ_d. Referring to Fig. 5 *[Fig. A3.5]*, the PLDG appears in the *X-Y* plane and rotates through angle Θ_R about the *Z*-axis. When this angle is zero, the grating lines of the PLDG are assumed parallel to the *X*-axis, whereupon the incident and diffracted rays of the grating satisfy the following equations:

$$\sin \Theta_{ix} + \sin \Theta_{dx} = \frac{\lambda}{d} \sin \Theta_R \qquad (3a)^1$$

$$\sin \Theta_{iy} + \sin \Theta_{dy} = \frac{\lambda}{d} \cos \Theta_R \qquad (3b)^1$$

where $\sin \Theta_{ix}$ and $\sin \Theta_{iy}$ are the components of the wave vector along the *X* and *Y* axes, respectively; $\sin \Theta_{dx}$ and $\sin \Theta_{dy}$ are the components of the diffracted wave vector along the *X* and *Y* axes, λ is the wavelength of the incident light, *d* is the grating period and Θ_R is the grating rotation angle. These equations indicate that diffraction from a rotating PLDG can be modeled as diffrac-

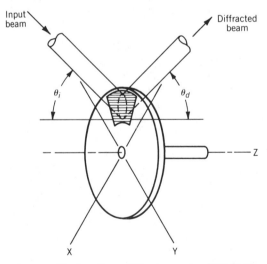

Figure A3.3. Disc substrate with plane linear diffraction grating (PLDG) illuminated at angle Θ_i and diffracting at angle Θ_d.

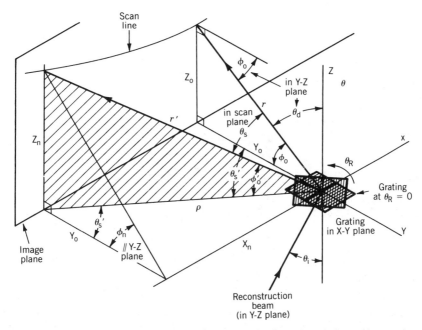

Figure A3.5. Coordinate system with parameters for analysis of scan bow during grating rotation.

tion from a two dimensional plane grating whose periods along orthogonal axes vary periodically in time from d to ∞, as illustrated in Figure 6 [Fig. A3.6]. The following notation conventions will be observed: subscripts i and d refer to the incident and diffracted waves, respectively; subscript o indicates the parameter is defined for the case of $\Theta_R = 0$; subscript n indicates the parameter is defined for $\Theta_R \neq 0$; Θ_s is the scan angle of the diffracted beam measured in the plane defined by the diffracted wave vector (and the Y-axis) and Θ_s' is the scan angle of the diffracted beam measured in the XY plane. Incident and diffracted angles for the grating are measured with regard to the grating normal (Z-axis). It will be assumed that the incident wave vector always lies in the YZ plane, so

$$\Theta_{ix} = 0 \quad \text{and} \quad \Theta_{iy} = \Theta_i. \qquad (3c)^{1}*$$

With reference to Figure 5 [Fig. A3.5], it is evident that if the diffraction angle Θ_d remains constant with scan angle, the scan line will bow upward as indicated in Figure 5. In order for the scan line to become straight, Θ_d must increase with scan angle. From the grating equation,

$$\sin \Theta_i + \sin \Theta_d = \lambda/d, \qquad (4)$$

*Refer to "Keyed Annotations."

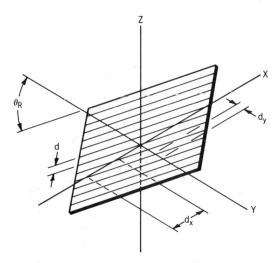

Figure A3.6. Effective grating spacings during grating rotation.

it is apparent that in order for Θ_d to increase, the incidence angle Θ_i must decrease. Insight into the relationship between apparent incidence angle and grating rotation angle can be achieved with the aid of Figure 7 *[Fig. A3.7]*. To illustrate how the apparent incidence angle changes as a function of rotation angle, the spinner in Figure 7 is depicted as being kept stationary while the incident beam is rotated about the z-axis. The projection of the rotated incident angle onto the YZ plane is:

$$\tan \Theta_i = \frac{y_1}{z_1} \cos \Theta_R. \tag{5}$$

The apparent incidence angle decreases from Θ_i to 0 as Θ_R increases from 0 to 90°. It will be shown that if Θ_d resides within a certain range, there exists a Θ_i value that minimizes the scan bow line.

The first step in deriving the conditions that minimize bow is to define parameters which are useful for characterizing the magnitude of the scan line bow. It is evident from Figure 5 that the change in the Z-coordinate of the scan line is an accurate measure of the bow. Since the change in the angle Φ_n is directly related to the change in the Z-coordinate, the bow can be expressed in terms of this variable. From the geometric relationships depicted in Figure 5 it is straight forward to show that:

$$\tan \Phi_n = \cot \Theta_d \sec \Theta_s'. \tag{6}^2$$

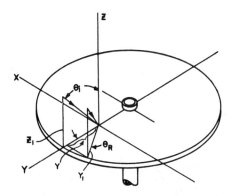

Figure A3.7. Relative change of incident angle Θ_i with effective rotation Θ_R.

For a bow free line, Φ_n is constant with regard to Θ_s'. To determine the conditions under which this occurs, Eq. (6) is differentiated with respect to Θ_s' while keeping Φ_n constant:

$$d\Theta_d = \sin \Theta_d \cos \Theta_d \tan \Theta_s' \, d\Theta_s'. \tag{7}^3$$

In order to solve Eq. (7), Θ_s' must be expressed in terms of Θ_R. By definition:

$$\tan \Theta_s' = \frac{x_d}{y_d} \tag{8}^4$$

where x_d and y_d are the components of the diffracted wave vector along the X and Y axis, respectively. Substituting from Eq. (3) gives:

$$\tan \Theta_s' = \frac{\tan \Theta_R}{1 - \dfrac{d}{\lambda} \dfrac{\sin \Theta_i}{\cos \Theta_R}}. \tag{9}$$

With the aid of Figure 5, one can show that:

$$\tan \Theta_s = \tan \Theta_s' \sin \Theta_d$$

$$= \frac{\tan \Theta_R \sin \Theta_d}{1 - \dfrac{d}{\lambda} \dfrac{\sin \Theta_i}{\cos \Theta_R}}. \tag{10}^5$$

Development for Eq. (11), which was not in original work, is introduced here with more information in the Keyed Annotations (superscript numerals). Since we are interested in solving Eq. (7) for the conditions where the PLDG is to be made ($\Theta_R = 0$), *the following approximation is utilized:*

$$d\Theta'_s = \frac{d\Theta_R}{1 - \dfrac{d}{\lambda} \dfrac{\sin \Theta_i}{\cos \Theta_R}}. \qquad (A)^6$$

Substituting Eqs. (9) and (A) into Eq. (7) gives:

$$d\Theta_d = \frac{\sin \Theta_d \cos \Theta_d \tan \Theta_R \, d\Theta_R}{\left[1 - \dfrac{d}{\lambda} \dfrac{\sin \Theta_i}{\cos \Theta_R} \right]^2}. \qquad (B)$$

To eliminate the differentials from this equation a second equation for Θ_d will be derived utilizing Eq. (3). By definition, the component of the diffracted wave vector lying in the plane of the grating is:

$$\sin \Theta_d = \sin \Theta_{dx} \hat{\imath} + \sin \Theta_{dy} \hat{\jmath}$$
$$= \left(\sin^2 \Theta_{dx} + \sin^2 \Theta_{dy} \right)^{1/2}. \qquad (C)$$

Substituting from Eq. (3) gives:

$$\sin \Theta_d = \left(\frac{\lambda^2}{d^2} - 2 \frac{\lambda}{d} \sin \Theta_i \cos \Theta_R + \sin^2 \Theta_i \right)^{1/2}. \qquad (D)^7$$

Differentiating this equation with respect to Θ_R gives:

$$d\Theta_d = \frac{(\lambda/d) \sin \Theta_i \sin \Theta_R \, d\Theta_R}{\cos \Theta_d \left(\dfrac{\lambda^2}{d^2} - 2 \dfrac{\lambda}{d} \sin \Theta_i \cos \Theta_R + \sin^2 \Theta_i \right)^{1/2}}. \qquad (E)^8$$

Substituting Eq. (E) into Eq. (B) and letting $\Theta_R \to 0$ gives:

$$\sin \Theta_i = \frac{\lambda}{d} \cos^2 \Theta_d \sec \Theta_R. \qquad (11)^8$$

Setting $\Theta_R = 0$ gives the making conditions for the PLDG:

$$\sin \Theta_i = \frac{\lambda}{d} - \frac{d}{\lambda}, \qquad (12a)^9$$

$$\sin \Theta_d = \frac{d}{\lambda}. \tag{12b}$$

These solutions depend on only the wavelength of light and the grating period. Real solutions for Θ_i and Θ_d exists only for the range where:

$$-1 \leq \frac{\lambda}{d} - \frac{d}{\lambda} \leq 1, \tag{13}$$

and

$$0 \leq \frac{d}{\lambda} \leq 1. \tag{14}$$

Since d can have only positive values, the maximum and minimum values that d can have are:

$$0.618\lambda \leq d \leq \lambda, \tag{15}$$

and

$$1 \leq \frac{\lambda}{d} \leq 1.618. \tag{16}$$

The maximum and minimum corresponding values of Θ_i and Θ_d are:

$$0° \leq \Theta_i \leq 89.445°$$

$$90° \geq \Theta_d \geq 38.17°. \tag{17}$$

There is a wide range of incident and diffracted angles over which this scan line bow minimization technique can be utilized, although certain angles are preferred.

If the PLDG is made and reconstructed according to Eqs. (12), the bow in the scan line is not eliminated but is minimized. In solving for Eqs. (12), Θ_R was set equal to zero, which insured that all the inflection points in the bowed scan line occurred at the center of the line ($\Theta_R = 0$). Figure 8 *[Fig. A3.8]* illustrates how the shape of the scan line changes as the incident angle is varied about the value predicted by Eq. (12a). The curve labeled Θ_i illustrates the scan line when the incident and diffracted angles satisfy Eqs. (12). The bow is essential zero for small rotation angles and monotonically increases with rotation

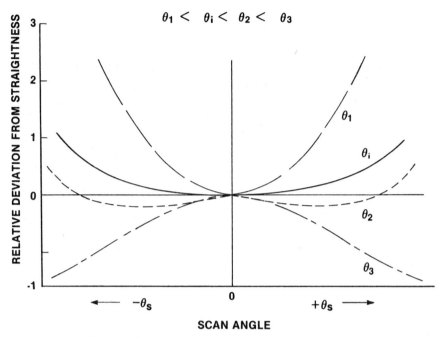

Figure A3.8. Scan line trajectories for different incident angles.

angle. For incident angles less than Θ_i, the monotonic increase in bow is faster. For incident angles larger than Θ_i, the scan line develops three inflection points as indicated by the curve labeled Θ_2. Since this curve first goes negative and then monotonically increases, its deviation from an ideal straight line can be made less than that of the Θ_i curve for a given rotation angle. For incidence angles much larger than Θ_i, the scan line begins to develop a fairly large negative bow as indicated by curve Θ_3.

To achieve the minimum bow in the scan line, the scanner would be set up to produce the Θ_2 curve of Figure 8 *[Fig. A3.8]*. Since the position of the inflection points of the scan line determine the maximum bow in the line, it is important to be able to predict the location of these points. Due to the differential approach utilized to solve this problem, information about the inflection points is contained in Eq. (11). By inspection of Eqs. (11) and (12) the following more generalized equations for the incident and diffraction angles are obtained.

$$\sin \Theta_i = \left(\frac{\lambda}{d} - \frac{d}{\lambda}\right) \sec \Theta_R \qquad (18a)$$

$$\sin \Theta_d = (1 - \sec \Theta_R) \frac{\lambda}{d} + \frac{d}{\lambda} \qquad (18b)$$

When the incident and diffracted angles satisfy Eqs. (18), the three inflection points occur at $\Theta_R = 0$, and $\pm \Theta_R$. In a typical scanner arrangement the inflection points are chosen to occur at 50–65% of the maximum rotation angle per scan. The following approximation is very useful for relating the grating rotation angle to the beam scanning angle [using Eq. (12b)]:

$$\Theta_s \cong \Theta_R / \sin \Theta_d = (\lambda/d) \Theta_R \qquad (19)$$

KEYED ANNOTATIONS

1. (*Eq. 3*):   From Eq. (3c), Eqs. (3a) and (3b) reduce to

$$\sin \Theta_{dx} = \frac{\lambda}{d} \sin \Theta_R \qquad (3a')$$

$$\sin \Theta_i + \sin \Theta_{dy} = \frac{\lambda}{d} \cos \Theta_R \qquad (3b')$$

2. (*Eq. 6*):   For $\Phi_o = \Phi'$ per Fig. A3.5,

$$\frac{Z_n}{\rho} = \frac{Z_o}{Y_o}$$

$$\frac{Z_n}{Y_n} = \frac{Z_o}{Y_o} \cdot \frac{\rho}{Y_o}$$

$$\tan \Phi_n = \cot \Theta_d \sec \Theta_s' \qquad (6)$$

Note: Typo in patent; Θ_n should be Φ_n.

3. (*Eq. 7*):   For Φ_n constant, write Eq. (6) $\equiv c$. Then

$$\cot \Theta_d = c \cos \Theta_s' \equiv f \qquad (a)$$

$$\frac{df}{d\Theta_d} = -\csc^2 \Theta_d \quad \text{and} \quad \frac{df}{d\Theta_s'} = -c \sin \Theta_s' \qquad (b)$$

$$\frac{d\Theta_d}{ds'} = c \frac{\sin \Theta_s'}{\csc^2 \Theta_d} = c \sin^2 \Theta_d \sin \Theta_s' \qquad (c)$$

substitution yields

$$c = \frac{\cos \Theta_d}{\sin \Theta_d} \sec \Theta_s' \qquad (d)$$

$$d\Theta_d = \sin \Theta_d \cos \Theta_d \tan \Theta_s' \, d\Theta_s' \qquad (7)$$

4. (*Eq. 8*):   Per Fig. A3.5,

$$\tan \Theta_s' = \frac{X_n}{Y_o} \equiv \frac{X_d}{Y_d}$$

5. (*Eq. 10*):   Per Fig. A3.5,

$$\tan \Theta_s = \frac{X_n}{r} = \frac{X_n}{Y_o} \frac{Y_o}{r}$$

$$= \tan \Theta_s' \sin \Theta_d$$

$$= \left[\text{Eq. (9)} \right] (\sin \Theta_d) \qquad (10)$$

Note: Typo in patent; denominator $\sin \Theta_R$ should read $\cos \Theta_R$.

6. (*Eq. (A)*):   From Eq. (9):

$$\frac{d\Theta_s'}{d\Theta_R} \approx \frac{\tan \Theta_s'}{\tan \Theta_R} \qquad \text{as } R \to 0$$

7. (*Eq. (D)*):   From Eq. (3) (letting $\lambda/d \equiv k$),

$$\sin^2 \Theta_{dx} = k^2 \sin^2 \Theta_R \qquad (a)$$

$$\sin^2 \Theta_{dy} = k^2 \cos^2 \Theta_R + \sin^2 \Theta_i - 2k \cos \Theta_R \sin \Theta_i \qquad (b)$$

$$(a) + (b) = k^2 - 2k \sin \Theta_i \cos \Theta_R + \sin^2 \Theta_i = (D)^2 \qquad (D)$$

8. (*Eqs. (E)*   From Eq. (E) for $\Theta_R \to 0$ ($\lambda/d \equiv k$),
   and 11):

$$\frac{d\Theta_d}{d\Theta_R} = \frac{k \sin \Theta_i \Theta_R}{\cos \Theta_d (k^2 - 2k \sin \Theta_i + \sin^2 \Theta_i)^{1/2}} \qquad (a)$$

in which the denominator $(\cdots)^{1/2} = k - \sin \Theta_i$. From Eq. (B) for $\Theta_R \to 0$,

$$\frac{d\Theta_d}{d\Theta_R} = \frac{\sin \Theta_d \cos \Theta_d \Theta_R}{(1 - \sin \Theta_i/k)^2} = \frac{k^2 \sin \Theta_d \cos \Theta_d \Theta_R}{(k - \sin \Theta_i)^2} \qquad (b)$$

Equating (a) and (b) gives us

$$\sin \Theta_i = k \cos^2 \Theta_d \left(\frac{\sin \Theta_d}{k - \sin \Theta_i} \right) \qquad (c)$$

Per Eq. (4), the term in parentheses $= 1$, whence

$$\sin \Theta_i = \frac{\lambda}{d} \cos^2 \Theta_d \qquad \text{for } \Theta_R \to 0 \qquad (11)$$

9. (*Eq. 12*):   From Eqs. (4) and (11), for $\Theta_R \to 0$,

$$\sin \Theta_i = k(1 - \sin^2 \Theta_d) = k - k(k - \sin \Theta_i)^2 \qquad (a)$$

$$= k - k^3 + 2k^2 \sin \Theta_i - k \sin^2 \Theta_i \qquad (b)$$

$$\sin^2 \Theta_i + \sin \Theta_i (1/k - 2k) + (k^2 - 1) = 0 \qquad (c)$$

whence, with real roots,

$$\sin \Theta_i = k - 1/k \qquad (12a)$$

and per Eq. (4),

$$\sin \Theta_d = 1/k \qquad (12b)$$

References

Short name	Reference
Agm	P. Agmon, A. C. Livanes, A. Katzir, and A. Yariv, "Simultaneous exposure and development of photoresist materials; an analytical model," *Appl. Opt.*, 16, 10, 2612–14 (Oct. 1977).
And	H. L. Anderson, Ed., *Physics Vade Mecum*, American Institute of Physics, New York, (1981), p. 251.
Ant	M. V. Antipin and N. G. Kiselev, "Laser beam deflector utilizing transmission holograms," *Tech. kino Telev.*, 6, 43–45 (June 1979) (in Russian). [See Appendix 2.]
Arn	S. M. Arnold, "Electron beam fabrication of computer generated holograms," *Opt. Eng.*, 24, 5, 803 (Sept./Oct. 1985).
Bai	H. S. Baird, U.S. Patent 1,962,474, "Scanning device for television," (June 12, 1934).
Bar 1	H. Bartelt and S. K. Case, "High efficiency hybrid computer-generated holograms," *Appl. Opt.*, 21, 16, 2886–90 (Aug. 15, 1982).
Bar 2	R. A. Bartolini, "Improved method for holograms recorded in photoresist," *Appl. Opt.*, 11, 5, 1275–76 (May 1972).
Bar 3	R. A. Bartolini, "Characteristics of phase relief holograms recorded in photoresist," *Appl. Opt.*, 13, 1, 129–39 (Jan. 1974).
Bed	L. Bedamian, "Acoustooptic laser recording," *Opt. Eng.*, 20, 1, 143–49 (Jan./Feb. 1981).
Bee	M. J. Beesley and J. G. Casteledine, "The use of photoresist as a holographic recording medium," *Appl. Opt.*, 9, 12, 2720–28 (Dec. 1970).
Bei 1	L. Beiser, U.S. Patent 3,614,193, "Light scanning system utilizing diffraction optics" (Oct. 19, 1971).
Bei 2	L. Beiser, "Optical scanning by means of rotating holograms," Abstract 1.4, IEEE/OSA Conf. Laser Eng. Appl. (May 30, 1973).

Short name	Reference
Bei 3	L. Beiser, "Diffraction light scanning systems using cylindrical and conic optics," private communication CBS Laboratories (June 25, 1973). [Cylindrical configuration published in Bei 7.]
Bei 4	L. Beiser, E. Darcy, and D. Kleinschmitt, "Holofacet laser scanning—an advanced data and image scanning technique," Proc. E-O Syst. Des. Conf., 75–81 (1973).
Bei 5	L. Beiser, D. Gabor, and D. Kleinschmitt, "Light scanning system utilizing flat diffraction gratings," private communication CBS Laboratories (Dec. 3, 1973).
Bei 6	L. Beiser, "Laser scanning systems," in *Laser Applications*, Vol. 2, M. Ross, Ed., Academic Press, New York (1974), pp. 55–159.
Bei 7	L. Beiser, "Advances in holofacet laser scanning," Proc. 1974 E-O Syst. Des. Conf., 333–40 (1975).
Bei 8	L. Beiser, "Holographic scan aberration correction: a clarification," *Appl. Opt.*, 16, 9, 2361 (Sept. 1977).
Bei 9	L. Beiser, Ed., *Laser Scanning and Recording, Selected Papers on*, SPIE Milestone Series, Vol. 378, Society of Photo-Optical Instrumentation Engineers, Bellingham, Wash. (1985).
Bei 10	L. Beiser, "Adaptation of holofacet scanning to reprographics and general data handling," *Proc. Soc. Photo Opt. Instrum. Eng.*, 299, 175–77 (1981).
Bei 11	L. Beiser, "Holographic scan with geometric and interferometric zone plates; a clarification," *Appl. Opt.*, 22, 11, 1613–15 (June 1, 1983).
Bei 12	L. Beiser, "Generalized equations for the resolution of laser scanners," *Appl. Opt.*, 22, 20, 3149–51 (Oct. 15, 1983).
Bei 13	L. Beiser, "Laser beam scanning for high density data extraction and storage," S.P.S.E. Symposium, Washington, D.C. (Oct. 1965).
Bei 14	L. Beiser, "Diffraction optics scanners," private communication, notes (May 17 and 18, 1969) and disclosure to CBS Laboratories (May 19 and 22, 1969).
Bei 15	L. Beiser and A. L. Dalisa, "Zone lens scanner quarterly report," private communication, CBS Laboratories (Mar. 31, 1969).
Bei 16	L. Beiser, E. Darcy, and D. Kleinschmitt, "Diffraction optics scanner for wideband film recording," RADC-TR-73-79 (Mar. 1973).
Bei 17	L. Beiser, "Holographic scanners for field-flattening systems," private communication, disclosure to CBS Laboratories (Oct. 16, 1972).
Bei 18	L. Beiser, "Diffraction optics scanning systems—transmission types," private communication, notes (May 24, 1969) and disclosure to CBS Laboratories (May 29, 1969)
Bjo	G. C. Bjorklund and G. T. Sincerbox, U.S. Patent 4,432,597, "Transmissive holographic optical element on aberrating substrate" (Feb. 21, 1984).
Bra 1	W. L. Bragg, "An optical method of representing the results of x-ray analysis," *Z. Krist.*, 70, 475–92 (1929).

Short name	Reference
Bra 2	A. Bramley, U.S. Patent 3,721,486, "Light scanning by interference grating and method" (Mar. 20, 1973).
Bra 3	R. E. Braiser, U.S. Patent 4,337,994, "Linear beam scanning apparatus especially suitable for recording data on light sensitive film" (July 6, 1982).
Bro	B. R. Brown and A. W. Lohmann, "Complex spatial filtering with binary masks," *Appl. Opt.*, 5, 967 (1966).
Bry 1	O. Bryngdahl, "Optical scanner-light deflection using computer-generated diffractive elements," *Opt. Commun.*, 15, 2, 237–40 (Oct. 1975).
Bry 2	O. Bryngdahl and W.-H. Lee, "Laser beam scanning using computer-generated holograms," *Appl. Opt.*, 15, 1, 183–94 (Jan. 1976).
Bry 3	O. Bryngdahl and W.-H. Lee, U.S. Patent 4,106,844, "Laser scanning system utilizing computer generated holograms" (Aug. 15, 1978).
Cam	D. K. Campbell and D. W. Sweeney, "Materials processing with CO_2 laser holographic scanner system," *Appl. Opt.*, 17, 23, 3727–37 (Dec. 1, 1978).
Cas 1	S. K. Case and W. J. Dallas, "Volume holograms constructed from computer-generated masks," *Appl. Opt.*, 17, 16, 2537–40 (Aug. 15, 1978).
Cas 2	S. K. Case and V. Gerbig, "Laser beam scanners constructed from volume holograms," *Opt. Eng.*, 19, 5, 711–15 (Sept./Oct. 1980).
Cas 3	S. K. Case and V. Gerbig, "Efficient and flexible laser beam scanners constructed from volume holograms," *Opt. Commun.*, 36, 2, 94–100 (Jan. 15, 1981).
Cat	R. T. Cato and L. D. Dickson, U.S. Patent 4,548,463, "Holographic scanner control based on monitored diffraction efficienty" (Oct. 22, 1985).
Cau	H. J. Caulfield, Ed., *Handbook of Optical Holography*, Academic Press, New York (1979).
Cha 1	E. B. Champagne, "Nonparaxial imaging magnification and aberration properties in holography," *J. Opt. Soc. Am.*, 57, 1, 51–55 (Jan. 1967).
Cha 2	E. B. Champagne and N. G. Massey, "Resolution in holography," *Appl. Opt.*, 8, 9, 1879–85 (Sept. 1969).
Cha 3	B. J. Chang and K. A. Winick, "Holographic optical elements," *Proc. Soc. Photo Opt. Instrum. Eng.*, 299, 157–62 (1981).
Cha 4	B. J. Chang and C. D. Leonard, "Dichromated gelatin for the fabrication of holographic optical elements," *Appl. Opt.*, 15, 2407 (1979).
Cha 5	B. J. Chang, "Dichromated gelatin holograms and their applications," *Opt. Eng.*, 19, 5, 642–48 (Sept./Oct. 1980).
Che 1	C. C. K. Cheng, U.S. Patent 4,224,509, "Holographic scan system" (Sept. 23, 1980).
Che 2	C. W. Chen, "Using conventional optical design program to design holographic optical elements," *Opt. Eng.* 19, 5, 649–53 (Sept./Oct. 1980).
Cin	I. Cindrich, "Image scanning by rotation of a hologram," *Appl. Opt.*, 6, 9, 1531–34 (Sept. 1967).

Short name	Reference
Clo 1	D. H. Close, "Optically recorded holographic elements," in *Handbook of Optical Holography*, H. J. Caulfield, Ed., Academic Press, New York (1979), Chap. 10.8, pp. 573–85.
Clo 2	D. H. Close and G. E. Moss, "Determination of ray directions in ray-tracing thick holograms," *J. Opt. Soc. Am.*, 63, 10, 1324 (Oct. 1973). [Paper abstract FE13, 1973 meeting OSA.]
Col	R. J. Collier, C. B. Burckhardt, and L. H. Lin, *Optical Holography*, 2nd ed., Academic Press, New York (1971).
Coo	D. J. Cook and A. A. Ward, "Reflection-hologram processing for high efficiency in silver-halide emulsions," *Appl. Opt.*, 23, 6, 934–41 (Mar. 15, 1984).
Dal	A. L. Dalisa, private communication to L. Beiser, CBS Laboratories (Dec. 12, 1968).
Dar	E. Darcy, "New high efficiency zone lens scanners," private communication to L. Beiser, CBS Laboratories (June 7, 1971).
Dic 1	L. D. Dickson and G. T. Sincerbox, "Optics and holography in the IBM supermarket scanner," *Proc. Soc. Photo Opt. Instrum. Eng.*, 299, 163–68 (Aug. 1981).
Dic 2	L. D. Dickson, U.S. Patent 4,416,505, "Method for making holographic optical elements with high diffraction efficiency" (Nov. 22, 1983).
Dic 3	L. D. Dickson, "Holographic scanning—principles and applications," *Proc. Soc. Photo Opt. Instrum. Eng.*, 498 (Aug. 1984).
Dog	D. E. Doggett, U.S. Patent 4,478,480, "Holographic spinner wobble correction system" (Oct. 23, 1984).
Edw 1	W. R. Edwards, "The Reflaxicon, a new reflective optical element and some applications," *Appl. Opt.*, 12, 8, 1940–45 (Aug. 1974).
Edw 2	W. R. Edwards, "Imaging properties of a conic axicon," *Appl. Opt.*, 13, 8, 1762–63 (Aug. 1974).
Eng	R. C. Enger and S. K. Case, "Optical elements with ultrahigh spatial frequency surface corregations," *Appl. Opt.*, 22, 20, 3220–28 (Oct. 25, 1983).
Fim	A. Fima, M. Pardo, and J. A. Quintana, "Improvement of image quality in bleached holograms," *Appl. Opt.*, 21, 9, 3412–13 (Oct. 1, 1982).
Fla 1	J. Flamand and G. Pieuchard, "High rate and resolution scanning possibilities with rotating stigmatic holographic gratings," Proc. E-O Syst. Des. Conf., 322–27 (1973).
Fla 2	J. Flamand, C. Malabry, A. Labeyrie, G. sur Yvette, and G. Pieuchard, U.S. Patent 3,628,849, "Diffraction gratings" (Dec. 21, 1971).
Fle	J. M. Fleischer, U.S. Patent 3,570,189, "Light scanning and printing system" (July 1973).
Fos	L. C. Foster, C. B. Crumly, and R. L. Cohoon, "A high-resolution linear optical scanner using a traveling wave acoustic lens," *Appl. Opt.*, 9, 2154 (Sept. 1970).

Short name	Reference
Fun 1	H. Funato, U.S. Patent 4,325,601, "Optical scanning device" (Apr. 20, 1982).
Fun 2	H. Funato, private communication.
Fun 3	H. Funato, "Holographic scanner for laser printer," *Proc. Soc. Photo Opt. Instrum. Eng.*, 390, 174–82 (Jan. 1983).
Gab 1	D. Gabor, "A new method of optical scanning," private communication, Report No. 9 to P. C. Goldmark, CBS Laboratories (Apr. 12, 1965).
Gab 2	D. Gabor, "Holographic scanning disc," private communication CBS Laboratories, Report No. 3 to P. C. Goldmark, CBS Laboratories (Apr. 3, 1967).
Gab 3	D. Gabor, "Holography, 1948–1972," *Proc. IEEE*, 60, 6, 655–68 (June 1972).
Ger 1	V. Gerbig, "Computer-holographic laser scanner," 1977/78 Annual report, Angewandte Optic Group, Physikal. Inst., Universitat Erlangen-Nürnberg, 31–33.
Ger 2	V. Gerbig, "Anamorphic holographic laser scanners and scan capacity," 1980 Annual report, Angewandte Optic Group, Physikal. Inst., Universitat Erlangen-Nürnberg, 16–17.
Ger 4	V. Gerbig, "Holografishe Lichtablenker" (Holographic light deflection) Ph.D. thesis, University of Erlangen-Nürnberg, (1981) (in German).
Ger 5	V. Gerbig, "Holographic scanning: a review," *Proc. Soc. Photo Opt. Instrum. Eng.*, 396, 28–35 (Apr. 1983).
Gou	J. Gould, "Best U.S. surveillance system reported moving to the middle east," *New York Times* (Thursday, Aug. 27, 1970), p. 1.
Gra	B. D. Grant, M. M. Hilden, C. R. Jones, and G. T. Sincerbox, U.S. Patent 4,422,713, "Method for making high efficiency holograms" (Dec. 27, 1983).
Har	P. Hariharan, "Holographic recording materials: recent developments," *Opt. Eng.*, 19, 5, 636–39 (Sept./Oct. 1980).
Hec	J. Hecker, H. Stern, and T. Heydenberg, U.S. Patent 4,573,758, "Beam deflection mechanism" (Mar. 4, 1986).
Hef	D. Heflinger, J. Kirk, R. Cordero, and G. Evans, "Submicron grating fabrication on GaAs by holographic exposure," *Opt. Eng.*, 21, 3, 537–41 (May/June 1982).
Her	R. P. Herloski, U.S. Patent 4,508,421, "Holographic scanning system utilizing a scan linearization lens" (Apr. 1985).
How	J. W. Howard, "Formulas for the coma and astigmatism of wedge prisms used in convergent light," *Appl. Opt.*, 24, 23, 4265–68 (Dec. 1, 1985).
Ih 1	C. S. Ih, U.S. Patent 3,953,105, "Holographic scanner utilizing auxiliary reflective surface" (Apr. 27, 1976).
Ih 2	C. S. Ih, "Holographic laser beam scanners utilizing an auxiliary reflector," *Appl. Opt.*, 16, 8, 2137–46 (Aug. 1977).

Short name	Reference
Ih 3	C. S. Ih and D. Kleinschmitt, "Holo-scanner technique for wideband recording," RADC-TR-74-181 (July 1974).
Ih 5	C. S. Ih, U.S. Patent 4,266,846, "Two dimensional scanners" (May 12, 1981).
Ih 6	C. S. Ih, E. LeDet, and N. S. Kopeika, "Characteristics of holographic scanners using a concave auxiliary reflector," *Appl. Opt.*, 20, 9, 1656–63 (May 1, 1981).
Ih 7	C. S. Ih and L.-Q. Xiang, "Compound holographic scanners," *Proc. Soc. Photo Opt. Instrum. Eng.*, 498, 191–8 (Aug. 1984).
Ike 1	H. Ikeda, M. Ando, and T. Inagaki, U.S. Patent 4,165,464, "Light scanning system" (Aug. 21, 1979).
Ike 2	H. Ikeda, M. Ando, and T. Inagaki, "Aberration correction for a P.O.S. hologram scanner," *Appl. Opt.*, 18, 13, 2166–70 (July 1979).
Ike 3	H. Ikeda, K. Yamazaki, F. Yamagishi, I. Sebata, and T. Inagaki, "Shallow-type truncated symbol-reading point-of-sale hologram scanner," *Appl. Opt.*, 24, 9, 1366–70 (May 1, 1985).
Ish 1	H. Ishikawa, U.S. Patent 4,480,892, "Light beam deflection apparatus" (Nov. 6, 1984).
Ish 2	Y. Ishii, "Reflection volume holographic scanners with field-curvature corrections," *Appl. Opt.*, 22, 22, 3491–99 (Nov. 15, 1983).
Ish 3	Y. Ishii and K. Murata, "Flat-field linearized scans with reflection dichromated gelatin holographic gratings," *Appl. Opt.*, 23, 12, 1999–2006 (June 15, 1984).
Ish 4	H. Ishikawa, U.S. Patent 4,571,020, "Hologram light deflector" (Feb. 18, 1986).
Iwa 1	F. Iwata and J. Tsujiuchi, "Characteristics of a photoresist and its replica," *Appl. Opt.*, 13, 6, 1327–36 (June 1974).
Iwa 2	H. Iwaoka and T. Shiozawa, "Aberration-free linear holographic scanner and its application to a diode laser printer," *Appl. Opt.*, 25, 1, 123–29 (Jan. 1, 1986).
Izz	L. L. Izzo and H. A. Cubberly, "Optical spot size study for data extraction from a transparency," AMRL-TR-65-175 (AD628588), Wright-Patterson Air Force Base, Ohio (1965).
Jen 1	C. F. Jenkins, U.S. Patent 1,679,086, "Spiral mounted lens disk" (July 31, 1928).
Jen 2	F. A. Jenkins and H. E. White, *Fundamentals of Optics*, 3rd ed., McGraw-Hill, New York (1957).
Job	Jobin-Yvon Optical Systems, *Diffraction Gratings, Ruled and Holographic*, Jobin-Yvon Optical Systems, Longjumeau, France (1973).
Joh 1	K. M. Johnson, L. Hesselink, and J. W. Goodman, "Holographic reciprocity law failure," *Appl. Opt.*, 23, 2, 218–27 (Jan. 15, 1984).

Short name	Reference
Joh 2	L. F. Johnson, "Optical writing of dense 1000-Å features in photoresist," *Appl. Opt.*, 21, 11, 1892–93 (Sept./Oct. 1980).
Jor	J. A. Jordan, Jr., P. M. Hirsch, L. B. Lesem, and D. L. Van Rooy, "Kinoform lenses," *Appl. Opt.*, 9, 8, 1883–87 (Aug. 1970).
Kay	D. B. Kay, U.S. Patent 4,428,643, "Optical scanning system with wavelength shift correction" (Jan. 31, 1984).
Kle 1	D. Kleinschmitt, "Polygonal substrate scanner utilizing diffraction gratings," private communication, CBS Laboratories (June 25, 1975).
Kle 2	D. Kleinschmitt, "Ellipsoidal and parabolic substrate diffractive light scanning systems," private communication, CBS Laboratories (July 3, 1975).
Kod	K. Kodate, H. Takenaka, and T. Kamiya, "Fabrication of high numerical aperture zone plates using deep ultraviolet lithography," *Appl. Opt.*, 23, 3, 504–7 (Feb. 1, 1984).
Kra 1	C. J. Kramer, U.S. Patent 4,067,639, "Holographic scanning spinner" (Jan 10, 1978)
Kra 2	C. J. Kramer, "Minimization of wobble effects in holographic scanners," private communication and presented 1979 CLEA, paper 17.9.
Kra 3	C. J. Kramer, U.S. Patent 4,239,326, "Holographic scanner for reconstructing a scanning light spot insensitive to mechanical wobble" (Dec. 16, 1980).
Kra 5	C. J. Kramer, U.S. Patent 4,243,293, "Holographic scanner insensitive to mechanical wobble" (Jan. 6, 1981).
Kra 6	C. J. Kramer, "Holographic laser scanners for nonimpact printing," *Laser Focus*, 17, 70–82 (June 1981).
Kra 7	C. J. Kramer, U.S. Patent 4,289,371, "Optical scanner using plane linear diffraction gratings on a rotating spinner" (Sept. 15, 1981).
Kra 8	C. J. Kramer, U.S. Patent 4,304,459, "Reflective holographic scanning system insensitive to spinner wobble effects" (Dec. 8, 1981).
Kra 9	C. J. Kramer, "Hologon laser scanners for nonimpact printing," *Proc. Soc. Photo Opt. Instrum. Eng.*, 390, 165–73 (1983).
Kra 10	C. J. Kramer, "Specification and acceptance test procedures for hologon laser scanner systems," Proc. S.P.I.E., 776, 81-104 (1987).
Kuo	W. A. Kuo, "Photoresist optical elements on spherical substrate for high resolution laser scanning applications," Ph.D. thesis, State University of New York at Stony Brook (Aug. 1985).
Kur	R. L. Kurtz and R. B. Owen, "Holographic recording materials—a review," *Opt. Eng.*, 14, 5, 393-401 (Sept./Oct. 1975).
Lat	J. N. Latta, "Computer-based analysis of hologram imagery and aberrations," *Appl. Opt.*, 10, 3, pt. L1, 599-608; pt. L2, 609-18 (Mar. 1971).
Lee 1	W.-H. Lee, "Holographic grating scanners with aberration correction," *Appl. Opt.*, 16, 5, 1392-99 (May 1977).
Lee 2	W.-H. Lee, "Computer-generated holograms: techniques and applications,"

Short name	Reference
	in *Progress in Optics*, Vol. 16, E. Wolf, Ed., North-Holland, New York (1979), pp. 120–32.
Leg	A. Legarcon and P. Chavel, "Le facteur de dualité N/τ dans les déflecteurs de faisceaux lasers: comparison entre déflecteurs holographiques et acousto-optiques," *Opt. Commun.*, 25, 2, 151–56 (May 1978).
Len	E. Lennemann, "Aerodynamic aspects of disk files," *IBM J. Res. Dev.*, 480–88 (Nov. 1974).
Lev	L. Levi, *Applied Optics*, Vol. 1, Wiley, New York (1968).
Lin	L. H. Lin and E. T. Doherty, "Efficient and aberration-free wavefront reconstruction from holograms illuminated at wavelengths differing from the forming wavelength," *Appl. Opt.*, 16, 6, 1314–18 (June 1971).
Loc	J. W. Locke and D. Mills, U.S. Patent 3,796,768, "Holographic image scanner/recorder system" (Mar. 5, 1974).
Loe	E. G. Loewen, M. Nevière, and D. Maystre, "Grating efficiency theory as it applies to blazed and holographic gratings," *Appl. Opt.*, 16, 10, 2711–21 (Oct. 1977).
Loh 1	A. W. Lohmann, "A new class of varifocal lenses," *Appl. Opt.*, 11, 7, 1669–71 (July 1970).
Loh 2	A. W. Lohmann and O. Bryngdahl, "A lateral wavefront shearing interferometer with variable shear," *Appl. Opt.*, 6, 11, 1934–37 (Nov. 1967).
Mar	G. F. Marshall, "Scanning devices and systems," in *Applied Optics and Optical Engineering*, Vol. 6, Academic Press, New York (1980), pp. 203–62.
Mat 1	R. M. Matic and E. W. Hansen, "Nondetour phase computer-generated holograms: an improved variation," *Appl. Opt.*, 21, 13, 2304–5 (July 1, 1982).
McL	J. H. McLeod, "The axicon: a new type of optical element," *J. Opt. Soc. Am.*, 44, 8, 592 (Aug. 1954).
McM 1	D. H. McMahon, A. R. Franklin, and J. B. Thaxter, "Light beam deflection using holographic scanning techniques," *Appl. Opt.*, 8, 2, 399–402 (Feb. 1969).
McM 2	D. H. McMahon, U.S. Patent 3,619,033, "Three-dimensional light beam scanner utilizing tandemly arranged diffraction gratings" (Nov. 9, 1971).
McM 4	R. H. McMann, Jr. "Optical scanner," patent application, CBS Laboratories, private communication (Nov. 24, 1967).
Meh	P. C. Mehta, K. S. S. Rao, and R. Hradaynath, "Higher order aberrations in holographic lenses," *Appl. Opt.*, 21, 24, 4553–58 (Dec. 15, 1982).
Mei 1	R. W. Meier, "Magnification and third-order aberration in holography," *J. Opt. Soc. Am.*, 55, 8, 987–92 (Aug. 1965).
Mei 2	R. W. Meier, "Depth of focus and depth of field in holography," *J. Opt. Soc. Am.*, 55, 12, 1693–94 (Dec. 1965).
Moh	M. G. Moharam and T. K. Gaylord, "Diffraction analysis of dielectric surface relief gratings," *J. Opt. Soc. Am.*, 72, 10, 1385–92 (Oct. 1982).

Short name	Reference
Nis	N. Nishihara, "Efficiency of a holographic wave-front converter," *Appl. Opt.*, 21, 11, 1995–2000 (June 1, 1982).
Nor	S. L. Norman and M. P. Singh, "Spectral sensitivity of Shipley AZ-1350J photoresist," *Appl. Opt.*, 14, 4, 818–20 (Apr. 1975).
Oli 1	J. Oliva, A. Fima, and J. A. Quintana, "Dichromated-gelatin holograms with improved signal-to-noise ratios," *Appl. Opt.*, 21, 16, 2891–93 (Aug. 15, 1982).
Oli 2	J. Oliva, P. G. Boj, and M. Pardo, "Dichromated gelatin holograms derived from Agfa 8E75HD plates," *Appl. Opt.*, 23, 2, 196–97 (Jan. 15, 1984).
Ono 1	Y. Ono, U.S. Patent 4,299,437, "Coherent beam scanner having a planar hologram illuminated by a convergent or divergent beam" (Nov. 10, 1981).
Ono 2	Y. Ono and N. Nishida, "Holographic laser scanner using generalized zone plates," *Appl. Opt.*, 21, 24, 4542–48 (Dec. 15, 1982).
Ono 3	Y. Ono and N. Nishida, "Holographic disk scanners for bow-free scanning," *Appl. Opt.*, 22, 14, 2132–36 (July 15, 1983).
Ono 4	Y. Ono and N. Nishida, "Holographic laser scanners for multi-directional scanning," *Appl. Opt.*, 22, 14, 2128–31 (July 15, 1983).
Ono 5	Y. Ono and N. Nishida, "Holographic zone plates for f-Θ and collimating lenses," *Appl. Opt.*, 25, 5, 794–97 (Mar. 1, 1986).
Pen	K. S. Pennington and J. S. Harper, "New phototechnology suitable for recording phase holograms and similar information in hardened gelatin," *Appl. Phys. Lett.*, 18, 3, 80–84 (Feb. 1, 1971).
Pie 1	G. Pieuchard, F. L. Fleury, J. Flamand, C. Malabry, A. Labeyrie, and G. sur Yvette, U.S. Patent 3,721,487, "Optical diffraction grating scanning device" (Mar. 20, 1973).
Pie 2	G. Pierattini and G. C. Righini, "Sensitometric and holographic data of Kodak 120-02 and Ilford HeNe2 plates," *Opt. Laser Technol.*, 72–74 (Apr. 1976).
Pol 1	R. V. Pole and H. P. Wollenmann, "Holographic laser beam deflector," *Appl. Opt.*, 14, 4, 976–80 (Apr. 1975).
Pol 2	R. V. Pole and H. W. Werlich, U.S. Patent 4,113,343, "Holographic opaque document scanner" (Sept. 12, 1978).
Pol 3	R. V. Pole, H. W. Werlich, and R. J. Krusche, "Holographic light deflection," *Appl. Opt.*, 17, 20, 3294–97 (Oct. 15, 1978).
Rad	B. M. Radl, U.S. Patent 4,284,994, "Laser beam recorder" (Aug. 18, 1981).
Ran	J. F. Rando, U.S. Patent 4,544,228, "Scanning method using a rotating prism" (Oct. 1, 1985).
Ral 1	R. Rallison and R. Lowe, "Hologram scanner design and fabrication in dichromated gelatin," *Proc. Soc. Photo Opt. Instrum. Eng.*, 353, 9 (Aug. 1982).
Ral 2	R. D. Rallison and G. L. Heidt, "Characteristics of dichromated gelatin scan-

Short name	Reference
	ners for printing applications," *Proc. Soc. Photo Opt. Instrum. Eng.*, 498, 199–203 (Aug. 1984).
Rig	A. K. Rigler, "Wavefront reconstruction by reflection," *J. Opt. Soc. Am.*, 55, 12, 1693 (Dec. 1965).
Rim	M. P. Rimmer, "Optical design for a holographic scanning system," *Proc. Soc. Photo Opt. Instrum. Eng.*, 103, 86 (1977).
Sch	P. C. Schubert, "Periodic image artifacts from continuous-tone laser scanners," *Appl. Opt.*, 25, 21, 3880–84 (Nov. 1, 1986).
She 1	N. K. Sheridon, "Production of blazed holograms," *Appl. Phys. Lett.*, 12, 9, 31618 (May 1, 1968).
She 2	N. K. Sheridon, U.S. Patent 3,580,657, "Blazed surface hologram" (May 25, 1971).
Shu	A. R. Shulman, *Optical Data Processing*, Wiley, New York (1970).
Sin	G. T. Sincerbox, "Holographic scanners: applications, performance, and design," in *Laser Scanning Using Mechanical Methods*, G. Marshall, Ed., Marcel Dekker, New York (1985).
Smi	H. M. Smith, *Principles of Holography*, Wiley, New York (1975).
Soa	O. D. D. Soares, "Review of resolution factors in holography," *Opt. Eng.*, 22, 4, SR-107–11 (July/Aug. 1983).
Sol 1	L. Solymar and D. J. Cooke, *Volume Holography and Volume Gratings*, Academic Press, New York (1981).
Sol 2	C. Solano, R. Lessard, and P. Roberge, "Red sensitivity of dichromated gelatin films," *Appl. Opt.*, 24, 8, 1189–92 (Apr. 15, 1985).
Ste	W. H. Stevenson, "Optical frequency shifting by means of a rotating diffraction grating," *Appl. Opt.*, 9, 3, 649–52 (Mar. 1970).
Str 1	G. W. Stroke, *An Introduction to Coherent Optics and Holography*, Academic Press, New York (1966).
Str 2	G. W. Stroke, "Stigmatism of concave holographic gratings in the case of wavelength shift," private communication, CBS Laboratories (Oct. 25, 1969).
Suh	T. Suhara, K. Kobayashi, H. Nishihara, and J. Kayama, "Graded index fresnel lenses for integrated optics," *Appl. Opt.*, 21, 11, 1966–71 (June 1, 1982).
Swe	W. C. Sweatt, "Describing holographic elements as lenses," *J. Opt. Soc. Am.*, 67, 6, 803–8 (June 1977).
Tai	A. M. Tai, "Two-dimensional image transmission through a single optical fiber by wavelength-time multiplexing," *Appl. Opt.*, 22, 23, 3826–32 (Dec. 1, 1983).
Toy	G. Toyen, "Generation of precision pixel clock in laser printers and scanners," *Proc. Soc. Photo Opt. Instrum. Eng.*, 84, 138–45 (Aug. 1976).

Short name	Reference
Urb 1	J. C. Urback and R. W. Meier, "Properties and limitations of hologram recording materials," *Appl. Opt.*, 8, 11, 2269–81 (Nov. 1969).
Urb 2	J. C. Urbach, T. S. Fisli, and G. K. Starkweather, "Laser scanning for electronic printing," *Proc. IEEE*, 70, 6, 597–618 (June 1982).
Whi	J. M. White, U.S. Patent 4,121,882, "Flat scan holographic laser beam deflector" (Oct. 24, 1978).
Wre	J. E. Wreede and A. Graube, U.S. Patent 4,329,409, "Process for fabricating stable holograms" (May 11, 1982).
Wya	J. C. Wyant, "Rotating diffraction grating laser beam scanner," *Appl. Opt.*, 14, 5, 1057–58 (May 1975).
Yok	K. Yokomori, "Dielectric surface-relief gratings with high diffraction efficiency," *Appl. Opt.*, 23, 14, 2303–10 (July 15, 1984).
You	M. Young, "Zone plates and their aberrations," *J. Opt. Soc. Am.*, 62, 8, 972–76 (Aug. 1972).
Zec	R. G. Zech, "Review of optical storage media," *Proc. Soc. Photo Opt. Instrum. Eng.*, 177, 56–66 (1979).
Zoo	J. D. Zook, "Light beam deflector performance: a comparative analysis," *Appl. Opt.*, 13, 4, 875–87 (Apr. 1974).

Index